Lecture Notes in Mathematics 1641

Editors:
A. Dold, Heidelberg
F. Takens, Groningen

T0220378

Springer
Berlin
Heidelberg
New York
Barcelona
Budapest
Hong Kong
London
Milan
Paris
Santa Clara
Singapore
Tokyo

Peter Abramenko

Twin Buildings and Applications to S-Arithmetic Groups

 Springer

Author

Peter Abramenko
Fachbereich Mathematik
Universität Frankfurt
Robert-Mayer-Str. 6–10
D-60054 Frankfurt, Germany
e-mail: abramenk@math.uni-frankfurt.de

Cataloging-in-Publication Data applied for

Die Deutsche Bibliothek - CIP-Einheitsaufnahme

Abramenko, Peter:
Twin buildings and applications to S-arithmetic groups / Peter
Abramenko. - Berlin ; Heidelberg ; New York ; Barcelona ;
Budapest ; Hong Kong ; London ; Milan ; Paris ; Santa Clara ;
Singapore ; Tokyo : Springer, 1996
  (Lecture notes in mathematics ; 1641)
  ISBN 3-540-61973-9
NE: GT

Mathematics Subject Classification (1991):
Primary: 20E42, 51E24, 20F32, 20G30, 20J05

Secondary: 05E25, 14L35, 51A50, 11E39, 51E12, 20F05, 20F55, 20G25,
              22E67, 17B67

ISSN 0075-8434
ISBN 3-540-61973-9 Springer-Verlag Berlin Heidelberg New York

Typesetting: Camera-ready TₑX output by the author
SPIN: 10520133   46/3142-543210 - Printed on acid-free paper

To my mother

and

to my brother

Preface

The present Lecture Notes volume combines aspects of two mathematical domains which are closely connected to each other: group theory and the theory of buildings. On the basis of investigations concerning twin buildings and subcomplexes of spherical buildings, finiteness properties of some S-arithmetic groups are derived (cf. the introduction for more details).

Large parts of this book are devoted to the theory of (twin) buildings and not only written with group theoretic applications in mind. The first two sections of Chapter I can serve very well as an introduction to twin buildings. §1 describes the group theoretic background of this new theory. The basic definitions and facts (see in particular Lemma 2) are introduced in §2. Though these results are mainly due to Tits, the complete proofs are given here since they are hard to find in the original papers or not yet published. The following two sections present some of my own investigations concerning twin buildings. These are applied at the end of Chapter I in order to yield Theorem A. This theorem is one major step on the way towards the results about S-arithmetic groups presented in Chapter III. The second main theorem needed in this context is proved in the course of Chapter II. It generalizes the well known Solomon–Tits theorem and states that certain subcomplexes of spherical buildings are homotopy equivalent to bouquets of spheres "in general" (cf. Section 4 of the introduction). The techniques of the proof combine Tits' classification of spherical buildings with some combinatorial ideas. This part of the book is accessible to every reader who has studied Tits' Lecture Notes volume on spherical buildings.

I gladly take the opportunity to thank at least a few of those who helped in one way or the other that this book could be written. First of all, I am greatly indebted to Prof. H. Behr for his personal interest in me, for the opportunities he offered to me and for many stimulating mathematical discussions. I am also very obliged to Prof. H. Abels who invited me several times to the SFB 343 in Bielefeld where I spent more than 17 months altogether and where parts of these notes first took shape. In this context, I would also like to thank the DFG for the financial support I received during that time. Last but not least I express my warmest thanks to Mrs. Ch. Belz for preparing the TeX-version of the present book and for doing this exceptionally well.

Frankfurt am Main, September 1996

Table of contents

Introduction

In the following pages, I will try to describe briefly the background of this book, the key questions, the progress that has been achieved and some of the problems which are left for future work.

1. Finiteness properties of S–arithmetic groups

Large parts of the present notes are devoted to the theory of buildings and are of interest in their own right. However, the origin of these investigations was a group theoretic question which I am going to describe now.

Since the last century, groups of invertible matrices have been studied extensively, partly because of their geometric significance (one may think of $O_m(\mathbb{R})$ and other "groups of motion"). Since the books of Weyl (1946) and Dieudonné (1948/55), important classes of these linear groups, namely the general and special linear, the orthogonal (often also the spin-), the symplectic and the unitary groups over skew fields, have been subsumed under the notion of "**classical groups**". A common feature of all classical groups is the fact that they can be defined by algebraic equations over a commutative subfield k of the skew field K in question if the latter is finite-dimensional over its center. In order to mention at least one example, I recall that any orthogonal group is of the form $O_m(k, Q) = \{g \in GL_m(k) \mid g^t Q g = Q\}$, showing that the entries g_{ij} of $g \in O_m(k, Q)$ satisfy a system of quadratic equations.

Starting with the papers of Borel and Chevalley in the mid 1950's, a systematic abstract theory of **linear algebraic groups** has been developed. Classical groups belong to the central subjects of this theory which was also strongly influenced by Lie theory on the other side. Chevalley's classification of semisimple algebraic groups over arbitrary algebraically closed fields, motivated by and at the same time vastly generalizing the corresponding result concerning semisimple complex Lie groups, represents one of the highlights in the theory of linear algebraic groups.

From the beginning, not only the Lie groups but also their arithmetic subgroups like e.g. $SL_m(\mathbb{Z})$, $Sp_{2m}(\mathbb{Z})$ have been of interest. By the way, "most" discrete subgroups of finite covolume in semisimple Lie groups are arithmetic by a celebrated theorem due to Margulis (for a precise statement and much more information about S-arithmetic groups, I refer to [M]). However, the notion of an arithmetic group in

1

its original meaning (involving only Q-groups and the Ring $\mathbb{Z}$) is too restrictive in many respects. As S-arithmetic subrings of global fields are natural generalizations of $\mathbb{Z}$, arithmetic groups are generalized by **S-arithmetic groups**. The prototype of an S-arithmetic group is represented by $\mathcal{G}(\mathcal{O}_S)$, where $\mathcal{G}$ is an algebraic group defined over a global field k and $\mathcal{O}_S$ is the ring of S-integers in k. In general, all subgroups of $\mathcal{G}(k)$ commensurable with $\mathcal{G}(\mathcal{O}_S)$ are called S-arithmetic. The applications mentioned in the title of this book refer to S-arithmetic groups of the form $\mathcal{G}(\mathbb{F}_q[t])$ or $\mathcal{G}(\mathbb{F}_q[t, t^{-1}])$, where $\mathcal{G}$ is a semisimple algebraic $\mathbb{F}_q$-group (cf. Chapter III, §2, Theorem C, Corollary 20 and Remark 17 iv)).

Regarding the structure of an S-arithmetic group Γ, some questions are suggesting themselves. Can one find a finite set of generators for Γ? Is Γ finitely presented? What about higher (homological) finiteness properties? In this context, I recall the following: Γ is said to be of **type $\mathbf{FP_n}$** ($n \in \mathbb{N}_0 \cup \{\infty\}$) if there exists a projective resolution of the trivial Γ-module $\mathbb{Z}$ such that the first $n + 1$ projective Γ-modules are finitely generated. This implies for example that all homology and cohomology groups $H_i(\Gamma)$, $H^i(\Gamma)$ are finitely generated abelian groups for $0 \le i \le n$. I mention in passing that commensurable groups are of the same FP-type. The properties FP_1 and finitely generated are equivalent; finite presentability implies FP_2 and "often" coincides with FP_2. (But there exist groups of type FP_2 which are not finitely presented as was shown recently, cf. [BB].) I refer to [Bi] for further interesting consequences of the FP_n-property. Modifying this notion slightly, one says that Γ is of **type $\mathbf{F_n}$** if there exists an Eilenberg–MacLane complex of type $K(\Gamma, 1)$ with finite n-skeleton (respectively, with finite m-skeleton for all $m \in \mathbb{N}$ if $n = \infty$). This is equivalent to requiring FP_n plus finite presentability in case $n \ge 2$ (cf. [Br1], Ch. VIII, §7).

As for answers to the questions stated above, one has to distinguish between the number field and the function field case. Suppose that Γ is an S-arithmetic subgroup of the linear algebraic group $\mathcal{G}$ defined over the global field k. We first assume that k is a number field. If Γ is arithmetic in the narrow sence (i.e. $k = \mathbb{Q}$ and $S = \{\infty\}$), it is always finitely presentable and even of type F_∞ according to results of Raghunathan (cf. [Ra]), respectively of Borel and Serre (cf. [BoS1]). If Γ is just S-arithmetic, a similar statement is not true any longer. For example, the additive group of $\mathbb{Z}[\frac{1}{p}]$ is not finitely generated. However, if $\mathcal{G}$ is **reductive**, Γ is again of type F_∞ as was shown by Borel and Serre in [BoS2]. In fact much more is proved there, for example that Γ is virtually of type FL and a duality group. Finally, the finitely presented S-arithmetic groups were completely characterized by Abels in the number field case (cf. [A1]).

Next we assume that k is a global function field, i.e. a finite extension of a rational function field $\mathbb{F}_q(t)$. We additionally suppose that $\mathcal{G}$ is reductive and isotropic over k (if $\mathcal{G}$ is k-anisotropic, then Γ is cocompact and hence of type F_∞ according to Theorem 4 of [Se1]). Contrary to the number field case, Γ need not even be of type F_1 here. It was first observed by Nagao in 1959 that $SL_2(\mathbb{F}_q[t])$ is not finitely generated (cf. [N]). Using group actions on trees, this fact was explained very nicely by Serre some years later (cf. [Se2], Ch. II, §1.6). Since 1959 several mathematicians contributed to the solution of the problem regarding finite generation and finite presentability of Γ (cf. the references in [Be1] and [Be2]). Eventually in 1992, Behr was able to give a full proof for the complete characterization — conjectured by him already years before — of all finitely presented S-arithmetic subgroups of reductive groups defined over global function fields (cf. [Be2]). The complete solution of the corresponding problem concerning higher finiteness properties of Γ will perhaps require another couple of decades. At the moment, the result is only known for some classes of S-arithmetic groups. It is shown in Stuhler's paper [Stu] that $SL_2(\mathcal{O}_S)$ is of type F_{s-1} but not of type FP_s for any S-arithmetic function ring $\mathcal{O}_S$ with $\#S = s$. (By the way, a similar result concerning the subgroup of all upper triangular matrices in $SL_2(\mathcal{O}_S)$ is derived in [Bu].) On the other side, $SL_{n+1}(\mathbb{F}_q[t])$ is of type F_{n-1} but not of type FP_n provided that q is "sufficiently big" (cf. [Ab1] and [A2]). Analogous results are derived — presupposing Theorem B (cf. Chapter II, §8), the proof of which is published here for the first time — in [Ab3] for all classical Chevalley groups over $\mathbb{F}_q[t]$. They will reappear as special cases of Theorem C below. Apart from Stuhler's paper and from the quantitatively slightly better result for $SL_{n+1}(\mathbb{F}_q[t])$ derived in [Ab1] (cf. Remark 17 ii)), this Theorem C contains all what is known at the moment concerning higher finiteness properties of S-arithmetic subgroups of reductive groups in the function field case.

2. Filtrations of Bruhat–Tits buildings

Almost all the results mentioned in the last paragraphs were proved by using topological methods. The definition of the property F_n already indicates that finiteness properties of groups are closely connected with topology. Even problems regarding finite generation and finite presentability, though in principle accessible to the methods of algebraic K-theory, are sometimes more successfully attacked by studying the action of the group in question on an appropriate topological space. This is well demonstrated by the proof of Behr's theorem given in [Be2].

Now for a given S-arithmetic subgroup Γ of a reductive group $\mathcal{G}$ defined over a global field k, a suitable space X with natural Γ-action can be obtained as follows: Denote by k_v the completion of k relative to v. Let X_v be the quotient space of $\mathcal{G}(k_v)$ modulo a maximal compact subgroup if v is archimedian, respectively the Bruhat–Tits building associated to $\mathcal{G}(k_v)$ as described in [BrT1,2] if v is non-archimedian. Then consider $X = \prod_{v \in S} X_v$ with diagonal Γ-action.

Though space and action enjoy "nice" properties (X is contractible and the Γ-action is proper), finiteness properties for Γ cannot be deduced directly unless the quotient X/Γ is compact. Essentially two methods have been applied so far in order to treat the non-cocompact case. The first consists in compactifying X/Γ suitably, thus yielding a compact $K(\Gamma, 1)$-complex. This idea was successfully exploited in the number field case (cf. [Ra] and [BoS1,2]), showing in particular that Γ is of type F_∞. A different approach has to be used if k is a function field. Most of the results in this case are based on an idea due to Stuhler. Studying $\Gamma = SL_2(\mathcal{O}_S)$, he filtered $X = \bigcup_{j \in \mathbb{N}_0} X_j$ by an increasing sequence of Γ-invariant subcomplexes

$X_0 \subseteq \ldots \subseteq X_j \subseteq X_{j+1} \subseteq \ldots$ with compact quotients X_j/Γ. Since the filtration constructed in [Stu] induces isomorphisms $\pi_i(X_j) \xrightarrow{\sim} \pi_i(X_{j+1})$ for all (sufficiently big) j and all $0 \leq i \leq s - 2$, all these homotopy groups are trivial in view of the contractibility of X. This implies that $SL_2(\mathcal{O}_S)$ is of type F_{s-1}. Using additionally a criterion due to Brown (cf. [Br2]), it is also easily deduced from the properties of the filtration that Γ is not of type F_{s-1} (Stuhler gave a different proof for this statement).

Stuhler's method was applied independently by Abels and me in order to determine the "finiteness length" (i.e. the maximal m such that Γ is of type F_m) of $\Gamma = SL_{n+1}(\mathbb{F}_q[t])$. Applying the "reduction theory" for X/Γ, one has many choices to construct filtrations of X which are finite modulo Γ. The problem is to verify the desired homotopy properties. The filtration used in [Ab1] yields a slightly better result (I refer again to Remark 17ii)) but the proof given in [A2] is more elegant and accessible to generalizations. It is Abel's filtration which will be applied in the present book (cf. Chapter I, §5).

3. Twin buildings

The action of $SL_{n+1}(\mathbb{F}_q[t])$ on the corresponding Bruhat–Tits building admits a simplicial fundamental domain in the strictest sense. More generally, given a (simply connected) Chevalley group $\mathcal{G}$, it was shown by Soulé in [So] that $X/\mathcal{G}(\mathbb{F}_q[t])$ can be identified with a "quartier" in the Bruhat–Tits building X associated to $\mathcal{G}(\mathbb{F}_q((t^{-1})))$. However, Soulé's proof ist not very transparent since it depends on calculations and not on geometric arguments.

A better understanding of this result is provided by the theory of **twin buildings**. The group $G = \mathcal{G}(\mathbb{F}_q[t, t^{-1}])$ possesses a **twin BN–pair** such that the two components Δ_+, Δ_- of the corresponding twin building are canonically isomorphic to the Bruhat–Tits buildings associated to $\mathcal{G}(\mathbb{F}_q((t^{-1})))$, $\mathcal{G}(\mathbb{F}_q((t)))$ (cf. Chapter I, §1, Example 3). $\mathcal{G}(\mathbb{F}_q[t])$ and $\mathcal{G}(\mathbb{F}_q[t^{-1}])$ are opposite maximal parabolic subgroups in G and are therefore stabilizers in G of two opposite vertices $0_- \in \Delta_-$ and $0_+ \in \Delta_+$. It follows (cf. Chapter I, §3, Proposition 3 and Corollaries 1,2) that the action of $\Gamma = \mathcal{G}(\mathbb{F}_q[t])$ on $X = \Delta_+$ admits the same simplicial fundamental domain as the action of $\mathcal{G}(\mathbb{F}_q[t^{-1}]) = \mathrm{Stab}_G\, 0_+$, namely a quartier in Δ_+.

Starting with this observation, it has turned out in many respects that the action of Γ on Δ_+ is better understood if one interprets $\Gamma = \mathcal{G}(\mathbb{F}_q[t])$ as the stabilizer of a vertex in Δ_-, the "twin" of Δ_+. Many arguments used in [Ab3] which I first thought to be dependent on specific features of Bruhat–Tits buildings can in fact be deduced more transparently in the framework of twin buildings (cf. in particular Ch. I, §5). At the same time, this approach admits more general results, for example concerning classical $\mathbb{F}_q$-groups instead of Chevalley groups over $\mathbb{F}_q[t]$ but also regarding certain Kac–Moody groups over $\mathbb{F}_q$.

Therefore, Chapter I is completely devoted to twin buildings. Motivated by the theory of Kac–Moody groups (cf. in particular [T8]), these objects which are generalizations of spherical buildings were introduced by Ronan and Tits. Roughly speaking, a twin building is a pair of buildings (Δ_+, Δ_-) together with an opposition relation between the chambers of Δ_+ and Δ_- possessing similar properties as the opposition relation in a spherical building. Only parts of what is known concerning twin buildings are published yet (cf. [T9-11] and [MR]; for the special case of twin trees see also [RT]). However, firstly the group theoretic background regarding twin BN-pairs, emphasizing the most important examples, and secondly the basic definitions and lemmata (which are either contained in [T9-11] or in [AR]) are recalled in the first two sections of Chapter I. §3 treats, as already mentioned, questions concerning fundamental domains for group actions on twin buildings. In order to deduce certain local properties of the filtration described in §5, one has to introduce "coprojections" in twin buildings. This is done in §4, the main result being Proposition 4, where coprojections are expressed by means of ordinary projections and the opposition relation. (In case the reader is interested in a suitable notion of "convexity" for twin buildings, I also refer to the appendix of §4.)

Finally, the goal of Chapter I, namely **Theorem A**, is deduced in §6. It states the following: Given a group G acting "strongly transitively" (cf. Definition 5 in §2) on a twin building (Δ_+, Δ_-) , where Δ_+, Δ_- are thick n-dimensional buildings, and a simplex $\emptyset \neq a_- \in \Delta_-$. Then the stabilizer G_{a_-} is of type F_{n-1} but not of type FP_n provided that certain conditions, namely (LF), (F) and (S), are satisfied. (LF) states that the apartments of Δ_+, Δ_- are infinite and locally finite which amounts to saying that they are either of irreducible affine or of compact hyperbolic type. (F) requires the finiteness of the intersections $G_{a_-} \cap G_{b_+}$ for all $\emptyset \neq b_+ \in \Delta_+$ and is equivalent to the finiteness of the ground field in most examples (cf. Corollary 7 in §6). The crucial condition (S) will be discussed below. As for applications, one should think of the example $G = \mathcal{G}(\mathbb{F}_q[t, t^{-1}])$, $a_- = 0_-$ and $G_{a_-} = \mathcal{G}(\mathbb{F}_q[t])$ described above. Another application is concerned with groups acting on twin trees (cf. Corollary 8 and Proposition 6) and generalizes the Nagao–Serre theorem. Further consequences of Theorem A will be stated below. But before I have to say a few words concerning (S).

4. Spherical subcomplexes of spherical buildings

It is usually difficult to determine the homotopy properties of a filtration $(X_j)_{j \in \mathbb{N}_0}$ directly. However, in [Stu], [A2], [Ab 1,3] and in [Be2], this problem could be reduced to questions concerning the **local** structure of the respective Γ-complex X . In all these cases, the isomorphisms $\pi_i(X_j) \xrightarrow{\sim} \pi_i(X_{j+1})$ were established up to a certain level of i by showing that the occurring "relative links" $\ell k_{X_{j+1}}(\sigma) \cap X_j$ have the "right" connectedness properties for all (poly-) simplices $\sigma \in X_{j+1} \setminus X_j$.

A similar proceeding is also possible with regard to the filtration described in Chapter I, §5, provided that the condition (LF) is satisfied. The latter implies that

6

the full links of non-void simplices in $X = \Delta_+$ are spherical buildings. Then the relative links with respect to the filtration are determined by Corollary 6 in §5. They are of the form $\Theta^0(a)$ with $a \in \Theta = \ell k_X(\sigma)$, where $\Theta^0(a)$ denotes the subcomplex of Θ generated by all chambers of Θ which contain a simplex opposite to a . Now the desired homotopy properties of the filtration of Δ_+ can be deduced from the following condition.

(S) If Θ is the full link of a non-void simplex in Δ_+ , then $\Theta^0(a)$ is (dim Θ)-spherical for any $a \in \Theta$.

Recall that (the geometric realization of) a d-dimensional simplicial complex is said to be **d-spherical** if it is $(d-1)$-connected. By the well known Solomon–Tits theorem, every spherical building Θ is (dim Θ)-spherical. Chapter II of the present book is devoted to the question whether the same is true for the subcomplexes $\Theta^0(a)$.

This question does not occur here for the first time. Already in connection with the determination of the finiteness length of $SL_{n+1}(\mathbb{F}_q[t])$, it was essential. It has also been investigated for other purposes than studying finiteness properties of groups. In [T7], Tits considers (among other things) the question whether $\Theta^0(c)$ is simply connected for a chamber c of an irreducible spherical rank 3 building Θ and translates it into a group theoretic problem (cf. also Chapter II, §2, Lemma 19). For finite rank 2 buildings, the connectedness of $\Theta^0(a)$ is investigated in [Brou]. By the way, two new results concerning generalized m-gons are outlined in Chapter II, §2, namely the Propositions 7 and 9. Proposition 7 can be used in order to verify in "almost all" cases the condition "(co)" introduced in [MR]. In that paper, an extension theorem for isometries between twin buildings is proved under the assumption (co) that $\Theta^0(c)$ is connected whenever Θ is a rank 2 link in one of the two components of the twin building and c is a chamber of Θ .

Unfortunately, it is definitely possible that $\Theta^0(a)$ is not spherical. To mention at least one example (others are discussed in §2 of Chapter II), I recall that $\Theta^0(c)$ is a torus if Θ is the A_3 building over $\mathbb{F}_2$ and c a chamber of Θ (cf. [T7], Section 16). Counter-examples of this type show that $\Theta^0(a)$ can only be expected to be spherical if Θ is "thick enough", i.e. if every panel ($:=$ codimension 1 face of a chamber) is contained in sufficiently many chambers. However, as Proposition 9 demonstrates, this does not suffice. Thus we are led to the following

Conjecture 1: Let Θ be a spherical Moufang building of rank $d+1$ which is "thick enough". Then $\Theta^0(a)$ is d-spherical for any $a \in \Theta$.

The proof of this conjecture for "classical buildings", i.e. for spherical buildings corresponding to classical groups (a definition not referring to groups is given in Chapter II, §3), occupies the largest part of Chapter II. The result is the following:

Theorem B: *Let Θ be a building of type A_{d+1}, C_{d+1} or D_{d+1} but not an exceptional C_3 building. Assume that every panel of Θ is contained in at least $(2^d + 1)$ chambers in the A_{d+1} case, respectively in at least $(2^{2d+1} + 1)$ chambers in the two other cases. Then $\Theta^0(a)$ is d-spherical for any $a \in \Theta$.*

The A_{d+1} case is considerably easier than the other two and was already established in [AA]. The general method underlying the proof of Theorem B is discussed in some detail in §3 of Chapter II. It should be applicable to buildings of exceptional type as well. However, the corresponding proofs will become technically complicated to such an extent that I have dispensed with trying to carry them out.

Some characteristic features of the proof of Theorem B are the following: One has to treat the spherical buildings case by case (this is already necessary for rank 2 Moufang buildings, cf. Proposition 7). In each case, one represents the buildings as flag complexes of certain geomtries and uses induction on the rank. In order to obtain sufficiently strong induction hypotheses, one also has to consider other subcomplexes than those of type $\Theta^0(a)$. It is one of the main difficulties (at least in the D_n case) to choose the "right" class of subcomplexes. Decreasing the rank of the buildings is connected with increasing the number of conditions defining the subcomplexes to be considered. One ends up with bounds as stated in the theorem though the complexes $\Theta^0(a)$ are probably already spherical under much milder assumptions. Apart from obvious quantative questions, there is also an interesting qualitative one: Is there a fixed constant $T \in \mathbb{N}$ such that Theorem B remains true after replacing $2^d + 1$, respectively $2^{2d+1} + 1$ by T ? The opinions about the answer to be expected diverge; my guess would be "no".

5. Group theoretic consequences

Equipped with Theorem B, it is now easy to draw conclusions from Theorem A. As already mentioned, the main application is concerned with certain S-arithmetic groups. For the first time, one also obtains some results regarding higher finiteness properties in the function field case where the linear algebraic groups are **non–split**. I will just state a variant — appearing as Corollary 20 in Chapter III, §2 — of the more detailed Theorem C of Chapter III here.

Theorem C': *Let $\mathcal{G}$ be an absolutely almost simple $\mathbb{F}_q$-group which is not of exceptional type. Denote by n the $\mathbb{F}_q$-rank of $\mathcal{G}$. Suppose $n \geq 1$ and $q \geq 2^{2n-1}$. Then $\mathcal{G}(\mathbb{F}_q[t])$ and $\mathcal{G}(\mathbb{F}_q[t, t^{-1}])$ are of type F_{n-1} , and $\mathcal{G}(\mathbb{F}_q[t])$ is not of type FP_n .*

For the restrictions occurring in this statement, Theorem B is responsible. But a similar result should also be true for the exceptional types.

Conjecture 2: The statement of Theorem C' holds for any absolutely almost simple $\mathbb{F}_q$-group of $\mathbb{F}_q$-rank $n \geq 1$ provided that q is "big enough".

I am careful about the cases with small q since I know meanwhile that Theorem A becomes definitely wrong if one cancels assumption (S) (cf. the last remark concerning (S) in Chapter I, §6, directly before Theorem A).

As far as $G = \mathcal{G}(\mathbb{F}_q[t, t^{-1}])$ is concerned, Theorem C' just represents a preliminary result since the action of G on the corresponding twin building is not fully exploited yet (cf. Chapter I, §6, Remark 7). I expect that the following is true:

Conjecture 3: Let $\mathcal{G}$ be an absolutely almost simple $\mathbb{F}_q$-group of $\mathbb{F}_q$-rank $n \geq 1$ and assume that q is "sufficiently big". Then $\mathcal{G}(\mathbb{F}_q[t, t^{-1}])$ is of type F_{2n-1} but not of type FP_{2n} .

Of course, much more general statements than the two conjectures mentioned above may be suspected in the function field case. However, since there are so few results, it does not seem to be appropriate at the moment to formulate these speculations explicitly.

Instead, I conclude this introduction by noting a further consequence of Theorem A which is obtained as a by-product. As already mentioned, the methods of Chapter I can also be applied to certain **Kac–Moody groups over $\mathbb{F}_q$**. From Theorem A (to be more precise: from the Corollaries 10 and 11) and from Proposition 11 in Chapter II, §2, one can deduce the following

Example: Let $\mathcal{G}_D$ be a Kac–Moody group functor as described in [T8] (cf. also Example 5 below). Assume that the Coxeter system (W, S) associated to D is of rank 4 and of compact hyperbolic type. Let q be a prime power ≥ 16. Then $G = \mathcal{G}_D(\mathbb{F}_q)$ is finitely presented. All proper parabolic (with respect to one of the two natural BN-pairs in G) subgroups of G are finitely presented but not of type FP_3.

I Groups acting on twin buildings

§ 1 Twin BN–pairs and RGD–systems

"BN–pairs", later on also called "Tits systems", were introduced by Tits in the context of linear algebraic groups. Tits extracted the BN–axioms from Chevalley's work (cf. in particular [C1] and [C2]) on semisimple groups and showed together with Borel that this axiomatization applies as well to arbitrary, not necessarily split reductive groups (cf. [BoT], §5, or [Bo], §21). BN–pairs have proved to be a powerful tool in group theory since, mainly for two reasons. Firstly, much is known of the structure of a group if it possesses a BN–pair (key-words: Bruhat decomposition, parabolic subgroups, criterions for simplicity). Secondly, a group with a BN–pair naturally acts on a simplicial complex, namely the building associated to it. This renders certain group theoretic problems accessible to geometric interpretations and solutions.

If a BN–pair belongs to the group $G = \mathcal{G}(k)$ of k-rational points of a reductive k-group $\mathcal{G}$ as described by Borel and Tits, it possesses some additional features due to the properties of the family $(\mathcal{U}_\alpha(k))_{\alpha \in \Phi}$ of unipotent subgroups associated to the (relative) root system $\Phi = {}_k\Phi$ of $\mathcal{G}$. These properties were axiomatized by Bruhat and Tits in [BrT1], §6.1, where they defined "root data" ("données radicielles"). The Tits system corresponding to a root datum always possesses a finite Weyl group because the latter coincides with the Weyl group of the root system indexing the root datum. Therefore, the associated building is of spherical type. In particular, it is possible to define when two chambers or two (minimal) parabolic subgroups are opposite. These opposition relations are among the important additional features of root data which have no analogues in the general theory of buildings and BN–pairs. Nevertheless, the notion of "oppositeness" can also be applied to certain situations where the Weyl groups are infinite. For example, every "minimal" Kac–Moody group G over a field gives rise to two BN–pairs (G, B_+, N) and (G, B_-, N) with the same Weyl group W (cf. [T8]). Though B_+ and B_- are not conjugate if W is infinite, they are related to each other in the same way as the opposite minimal parabolic subgroups B and $w_0 B w_0^{-1}$ of a BN–pair with finite Weyl group, where w_0 denotes the element of maximal length of the latter. A precise formulation of the relationship between B_+ and B_- leads to the axioms of a "twin BN–pair" which will be recalled below. The geometric structures corresponding to twin BN–pairs are "twin buildings". They were introduced by Ronan and Tits (cf. [T9], [T11] and [RT]) and will be treated in

11

detail in the following sections.

As BN–pairs with finite Weyl groups are generalized by twin BN–pairs, root data are generalized by "RGD–systems" (cf. [T11], §3.3; RGD = Root Groups Data). Many twin BN–pairs, among them all those belonging to Kac–Moody groups, result from RGD–systems, and the axioms for the root groups are often easily verified. For this reason RGD–systems will also be introduced below. The geometric significance of these systems, namely the Moufang property of the associated twin buildings, will play hardly any role in the following.

For the basic definitions and properties of Coxeter systems and Tits systems, the reader is referred to [Bou2], Ch. IV. Throughout this book, the following notations will be used:

I denotes a finite index set and $\mathbf{M} = (m_{ij})_{i,j \in I}$ a **Coxeter matrix**, i.e. a symmetric matrix satisfying $m_{ij} \in \mathbb{N} \cup \{\infty\}$ for all $i, j \in I$, $m_{ii} = 1$ for all $i \in I$ and $m_{ij} \geq 2$ for all $i \neq j$. A Coxeter system (W, S) is said to be **of type** M if W possesses a presentation of the form

$$W = \langle s_i, i \in I \mid (s_i s_j)^{m_{ij}} = 1 \text{ for all } i, j \in I \text{ satisfying } m_{ij} \neq \infty \rangle$$

with $\{s_i \mid i \in I\} = S$. Every Coxeter system with finite S is of type M for suitable I and M . We set $W_J := \langle s_i \mid i \in J \rangle \leq W$ for every subset J of I and denote by $\ell = \ell_S : W \longrightarrow \mathbb{N}_0$ the usual length function with respect to S . A Tits system (G, B, N, S) is called **of type** M if the corresponding Coxeter system $(N/(B \cap N), S)$ is of type M.

Definition 1 (cf. [T11], §3.2):   *Let (G, B_+, N, S) and (G, B_-, N, S) be two Tits systems (of type M) with the same Weyl group $W = N/(B_+ \cap N) = N/(B_- \cap N)$. Then (G, B_+, B_-, N, S) is called a* **twin BN–pair** *(of type M) if the following conditions are satisfied:*

(TBN1) $B_\varepsilon w B_{-\varepsilon} s B_{-\varepsilon} = B_\varepsilon w s B_{-\varepsilon}$ for $\varepsilon \in \{+, -\}$ and all $w \in W$, $s \in S$ such that $\ell(ws) < \ell(w)$

(TBN2) $B_+ s \cap B_- = \emptyset$ for alle $s \in S$

Example 1:   Let (G, B, N, S) be a Tits system with finite Weyl group $W = N/(B \cap N)$. Denote by w_0 the element of maximal length of W. Then

12

$(G, B, w_0 B w_0^{-1}, N, S)$ is a twin BN–pair.

Remark 1: As we shall see in the proof of Lemma 1 (cf. also [T11], §3.2), any twin BN–pair (G, B_+, B_-, N, S) satisfies the following two axioms:

(TBN1)′ $B_\varepsilon w B_{-\varepsilon} s B_{-\varepsilon} \subseteq B_\varepsilon \{w, ws\} B_{-\varepsilon}$ for $\varepsilon \in \{+, -\}$, $w \in W$ and $s \in S$

(TBN2)′ $B_+ w \cap B_- = \emptyset$ for all $w \in W \setminus \{1\}$

Note that (TBN1)′ and (TBN2)′ are also satisfied by (G, B, B, N, S) for every Tits system (G, B, N, S) and are therefore weaker than (TBN1) and (TBN2). Nevertheless, (TBN1)′ and (TBN2)′ would be sufficient for our purposes during the first three paragraphs as follows from the results of [Ab5] concerning "pre-twin BN-pairs" and "pre-twin buildings".

For a twin BN–pair, there exists a decomposition analogous to the Bruhat decomposition but involving B_+ and B_- at the same time. This "Birkhoff decomposition" is derived in [T11], §3.2, but will be proved anew below for the convenience of the reader.

Lemma 1: *Let (G, B_+, B_-, N, S) be a twin BN-pair with Weyl group W. Then, for $\varepsilon \in \{+, -\}$, the map*

$$\beta_\varepsilon : W \longrightarrow B_\varepsilon \setminus G / B_{-\varepsilon} \, , \quad w \longmapsto B_\varepsilon w B_{-\varepsilon}$$

is bijective.

Proof: We first derive the axioms (TBN1)′ and (TBN2)′ mentioned in Remark 1.

(TBN1)′: In view of (TBN1), it suffices to consider the case $\ell(ws) > \ell(w)$. Applying (TBN1) to $w' = ws$ and recalling the identity $B_{-\varepsilon} s B_{-\varepsilon} s B_{-\varepsilon} = B_{-\varepsilon} \cup B_{-\varepsilon} s B_{-\varepsilon}$, we obtain

$$\begin{aligned}
(1) \quad B_\varepsilon w B_{-\varepsilon} s B_{-\varepsilon} &= (B_\varepsilon w s B_{-\varepsilon} s B_{-\varepsilon}) s B_{-\varepsilon} \\
&= B_\varepsilon w s (B_{-\varepsilon} \cup B_{-\varepsilon} s B_{-\varepsilon}) = B_\varepsilon \{w, ws\} B_{-\varepsilon}
\end{aligned}$$

(TBN2)′: Assume $B_+ w \cap B_- \neq \emptyset$ and hence $B_+ w B_- = B_+ B_-$ for $w \in W \setminus \{1\}$. Choose $s \in S$ such that $\ell(ws) < \ell(w)$. Applying (TBN1), we obtain $B_+ w s B_- = (B_+ w B_-) s B_- = B_+ B_- s B_-$. Spezializing (1), this yields $B_+ w s B_- = B_+ B_- \cup B_+ s B_-$. Therefore $B_+ B_- = B_+ s B_- \, (= B_+ w s B_-)$. But then

13

$B_+ s \cap B_- \neq \emptyset$, contradicting (TBN2).

Now (TBN1)$'$ implies, by induction on $\ell(w')$, $\quad B_\varepsilon w B_{-\varepsilon} w' B_{-\varepsilon} \subseteq B_\varepsilon W B_{-\varepsilon}$ for all $w, w' \in W$. Combining this with the Bruhat decomposition, we obtain $G = B_{-\varepsilon} W B_{-\varepsilon} = B_\varepsilon B_{-\varepsilon} W B_{-\varepsilon} = B_\varepsilon W B_{-\varepsilon}$. Hence β_ε ist surjective.

To show that β_ε is also injective, we deduce from $B_\varepsilon w' B_{-\varepsilon} = B_\varepsilon w B_{-\varepsilon}$ for $w, w' \in W$ by induction on $\min\{\ell(w), \ell(w')\}$ that $w' = w$. We may assume $\ell(w') \leq \ell(w)$. The case $\ell(w') = 0$ is settled by (TBN2)$'$. Now assume $w' \neq 1$ and choose $s \in S$ such that $\ell(w's) < \ell(w')$. From (TBN1)$'$, we deduce $B_\varepsilon w' s B_{-\varepsilon} \subseteq B_\varepsilon w' B_{-\varepsilon} s B_{-\varepsilon} = B_\varepsilon w B_{-\varepsilon} s B_{-\varepsilon} \subseteq B_\varepsilon \{w, ws\} B_{-\varepsilon}$ and therefore $B_\varepsilon w' s B_{-\varepsilon} = B_\varepsilon w B_{-\varepsilon}$ or $B_\varepsilon w' s B_{-\varepsilon} = B_\varepsilon w s B_{-\varepsilon}$. In view of $\ell(w's) < \ell(w') \leq \ell(w)$ and the induction hypothesis, this implies $w's = ws$ and hence $w' = w$. $\qquad\square$

Before defining RGD-systems now, we have to recall how certain notions originally connected with root stystems can be generalized. For this purpose, we shall use some basic facts concerning Coxeter complexes which may be found for example in [T1] §2, or in [Br3], Ch. III. To every Coxeter system (W, S), there is naturally associated a Coxeter complex $\Sigma = \Sigma(W, S)$. The chambers of Σ are identified with the elements of W , and W coincides with the group of type-preserving automorphisms of Σ . A **"root"** of Σ is the image of a folding. If α is a root, the opposite root, i.e. the image of the opposite folding, will be denoted by $-\alpha$. A root is called **"positive"** if it contains 1 and **"negative"** otherwise. If $S = \{s_i \,|\, i \in I\}$, we denote by $\alpha_i \, (i \in I)$ the unique root containing 1 but not s_i . The set of chambers of α_i is equal to $\{w \in W \,|\, \ell(s_i w) > \ell(w)\}$. Set

$$
\begin{aligned}
\Phi &:= \{\alpha \,|\, \alpha \text{ is a root of } \Sigma\} = \{w\alpha_i \,|\, w \in W \text{ and } i \in I\} \\
\Phi_+ &:= \{\alpha \in \Phi \,|\, \alpha \text{ is positive}\} = \{w\alpha_i \,|\, w \in W, i \in I \text{ and } \ell(ws_i) > \ell(w)\} \\
\Phi_- &:= \{\alpha \in \Phi \,|\, \alpha \text{ is negative}\} = \{-\alpha \,|\, \alpha \in \Phi_+\}
\end{aligned}
$$

Motivated by the theory of Kac–Moody algebras, a pair of roots is called **prenilpotent** if $\alpha \cap \beta$ as well as $(-\alpha) \cap (-\beta)$ contains at least one chamber (cf. [T8], §§ 3.2, 3.4 and 5.1). For every prenilpotent pair $\{\alpha, \beta\}$, we set

$$
\begin{aligned}
[\alpha, \beta] &:= \{\gamma \in \Phi \,|\, \alpha \cap \beta \subseteq \gamma \text{ and } (-\alpha) \cap (-\beta) \subseteq (-\gamma)\} \text{ and} \\
(\alpha, \beta) &:= [\alpha, \beta] \setminus \{\alpha, \beta\} .
\end{aligned}
$$

Definition 2: *Let* (W, S), Φ, Φ_+ *and* Φ_- *be as above. A triple* $(G, (U_\alpha)_{\alpha \in \Phi}, H)$

consisting of a group G , a family of subgroups $(U_\alpha)_{\alpha \in \Phi}$ and a subgroup $H \leq G$ normalizing each $U_\alpha\,(\alpha \in \Phi)$ is called an **RGD-system** *if it satisfies the following conditions:*

(RGD 0) $U_\alpha \neq \{1\}$ for all $\alpha \in \Phi$

(RGD 1) For each prenilpotent pair $\{\alpha, \beta\} \subset \Phi$ with $\alpha \neq \beta$, the commutator $[U_\alpha, U_\beta]$ is contained in $\langle U_\gamma \,|\, \gamma \in (\alpha, \beta)\rangle$

(RGD 2) Given $i \in I$ and $u \in U_{\alpha_i} \setminus \{1\}$, there exists an $m(u) \in U_{-\alpha_i} u U_{-\alpha_i}$ satisfying $m(u)U_\alpha m(u)^{-1} = U_{s_i \alpha}$ for all $\alpha \in \Phi$. Furthermore, $m(u)H = m(u')H$ for any two $u, u' \in U_{\alpha_i} \setminus \{1\}$

(RGD 3) $HU_+ \cap U_- = \{1\}$ if $U_\varepsilon := \langle U_\alpha \,|\, \alpha \in \Phi_\varepsilon \rangle$ for $\varepsilon \in \{+, -\}$

(RGD 4) $G = H\langle U_\alpha \,|\, \alpha \in \Phi \rangle$

Remark 2:   The axioms stated above are not identical with but equivalent to those formulated by Tits in [T11], §3.3. Because I am mainly interested in conditions which imply the TBN-axioms and are easily verified in concrete examples, I made two changes (cf. also the hints in [Ab5], proof of Lemma 5):

1) Tits defined H to be $\bigcap\limits_{\alpha \in \Phi} N_G(U_\alpha)$. Since it is often tedious to calculate this intersection directly, I only required $H \subseteq \bigcap\limits_{\alpha \in \Phi} N_G(U_\alpha)$. But it is easy to show that the conditions stated in Definition 2 actually imply $H = \bigcap\limits_{\alpha \in \Phi} N_G(U_\alpha)$.

2) Instead of the above (RGD3), Tits only required for each $i \in I$ that $U_{-\alpha_i}$ should not be contained in U_+ . However, by the methods developed in [T8], Section 5 (cf. in particular the proof of Theorem 2), first the equation "$HU_+ \cap HU_- = H$" and then also our (RGD3) can be deduced from Tits' weaker condition and the other RGD-axioms. Because it is not difficult to verify $HU_+ \cap U_- = \{1\}$ directly in most of the examples I am interested in, I dispense with appealing to Tits' tricky arguments here. This has the advantage that the proof of the following proposition becomes more elementary. By the way, Tits' remark in [T11], §3.3, that this proof "is less easy than the familar proof in the special case where W is finite" exactly refers to the less elementary arguments which are necessary if one works with his weaker version of axiom (RGD3).

15

Proposition 1 (cf. [T11], §3.3, Propostion 4): *If* $(G, (U_\alpha)_{\alpha \in \Phi}, H)$ *is an RGD-system with Coxeter system* (W, S) *of type* M *and*
$N := \langle H, m(u) \mid u \in U_{\alpha_i} \setminus \{1\}, i \in I \rangle$, *then* (G, HU_+, HU_-, N, S) *is a twin BN-pair of type* M .

Proof: Set $B_\alpha := HU_\alpha$ ($\alpha \in \Phi$), $B_+ := HU_+$, $B_- := HU_-$ and verify the axioms (RD1) – (RD5) of [T8], Section 5, for the triple $(G, (B_\alpha)_{\alpha \in \Phi}, H)$. Then $N/B_+ \cap N = N/B_- \cap N = N/H$ is isomorphic to W and (G, B_ε, N, S) is a Tits system for $\varepsilon \in \{+, -\}$. Furthermore, (G, B_+, B_-, N, S) satisfies (TBN1). All these statements are proved in loc. cit. by a similar reasoning as in [BrT1], §6.1, where root data are treated. It is more difficult to deduce $B_+ \cap B_- = H$ from the RD-axioms. Here, the tricky arguments mentioned in Remark 2 have to be used. But $B_+ \cap B_- = H$ is of course a direct consequence of our axiom (RGD3). Having established this equality, (TBN2) is easily derived as well (cf. [T8], Corollary 1 of Theorem 2). □

Important examples of RGD-systems belong to Coxeter systems of **affine Weyl groups**. In these cases, the notions introduced before Definition 2 specialize as follows: Let Ψ be a reduced and irreducible root system in the Euclidean space $V = \mathbb{R}^n$, $\Pi = \{a_1, \ldots, a_n\}$ a base of Ψ, Ψ_+ the corresponding system of positive roots (in the classical sense), $\Psi_- = -\Psi_+$ and a_0 the root of maximal height in Ψ_+ . Denote by $L_{a,\ell}$ ($a \in \Psi$, $\ell \in \mathbb{Z}$) the hyperplane $L_{a,\ell} := \{v \in V \mid (a, v) + \ell = 0\}$, by $s_{a,\ell}$ the reflection of V fixing $L_{a,\ell}$ pointwise and by $W = W_{\text{aff}}(\Psi)$ the group of affine transformations of V generated by $\{s_{a,\ell} \mid a \in \Psi, \ell \in \mathbb{Z}\}$.
Set $I := \{i \in \mathbb{N}_0 \mid 0 \leq i \leq n\}$, $s_0 := s_{-a_0,1}$, $s_i := s_{a_i,0}$ ($1 \leq i \leq n$) and $S := \{s_i \mid i \in I\}$. Then (W, S) is a Coxeter system. The associated Coxeter complex $\Sigma = \Sigma(W, S)$ is isomorphic to the simplicial complex obtained from the decomposition of V into cells induced by $\mathcal{H} := \{L_{a,\ell} \mid a \in \Psi, \ell \in \mathbb{Z}\}$. All these statements are well known, cf. for example [Bou2], Ch. V, §3, and Ch. VI, §2, or [Br3], Ch. VI, §1.

The roots of Σ can be identified with the half spaces of V bounded by elements of $\mathcal{H}$. Setting $\alpha_{a,\ell} := \{v \in V \mid (a, v) + \ell \geq 0\}$, we therefore obtain $\Phi = \{\alpha_{a,\ell} \mid a \in \Psi, \ell \in \mathbb{Z}\}$. The chamber called 1 in the group theoretic description of Σ becomes the open cell $c_0 := \{v \in V \mid (a_0, v) < 1 \text{ and } (a_i, v) > 0 \text{ for all } 1 \leq i \leq n\}$ in the geometric description. The roots containing c_0 but not $s_i c_0$ ($i \in I$) are $\alpha_0 := \alpha_{-a_0,1}$ and

$\alpha_i := \alpha_{a,0}$ for $1 \leq i \leq n$. Furthermore,

$$\Phi_+ = \{\alpha_{a,\ell} \in \Phi \mid (a \in \Psi_+ \text{ and } \ell \geq 0) \text{ or } (a \in \Psi_- \text{ and } \ell \geq 1)\} \text{ and}$$
$$\Phi_- = \{\alpha_{a,\ell} \in \Phi \mid (a \in \Psi_+ \text{ and } \ell \leq -1) \text{ or } (a \in \Psi_- \text{ and } \ell \leq 0)\}$$

Finally, a pair of roots $\{\alpha_{a,\ell}, \alpha_{b,m}\}$ is prenilpotent if and only if $b \neq -a$. It is easy to check that

$$[\alpha_{a,\ell}, \alpha_{b,m}] = \{\alpha_{pa+qb, p\ell+qm} \mid p, q \geq 0, pa + qb \in \Psi \text{ and } p\ell + qm \in \mathbb{Z}\}$$

in this case.

In the following examples, k denotes a field and $k[t, t^{-1}]$ the ring of Laurent polynomials in t over k.

Example 2: The group $G = SL_{n+1}(k[t, t^{-1}])$ We shall show that G possesses an RGD–system with Coxeter group $W = W_{\text{aff}}(A_n)$. Denote by $\varepsilon_1, \ldots, \varepsilon_{n+1}$ the canonical basis of $\mathbb{R}^{n+1}$ and set
$\Psi = \{a_{ij} := \varepsilon_i - \varepsilon_j \mid 1 \leq i \neq j \leq n + 1\}$. This is a root system of type A_n in
$V = \left\{ \sum_{i=1}^{n+1} \lambda_i \varepsilon_i \in \mathbb{R}^{n+1} \mid \lambda_1 + \ldots + \lambda_{n+1} = 0 \right\}$.
Choose $\Pi = \{a_i := a_{i\,i+1} \mid 1 \leq i \leq n\}$ as base of Ψ .
We denote by $\begin{pmatrix} a & b \\ c & d \end{pmatrix}_{ij}$ the $(n + 1) \times (n + 1)$-matrix (g_{rs}) with entries
$g_{ii} = a$, $g_{ij} = b$, $g_{ji} = c$, $g_{jj} = d$, $g_{rr} = 1$ for all $r \neq i, j$ and $g_{rs} = 0$ everywhere else.
Set

$$e_{ij}(\lambda) := \begin{pmatrix} 1 & \lambda \\ 0 & 1 \end{pmatrix}_{ij} \quad (1 \leq i \neq j \leq n + 1),$$
$$U_\alpha := \{e_{ij}(ct^{-\ell}) \mid c \in k\} \quad \text{for} \quad \alpha = \alpha_{a_{ij}, \ell} \in \Phi = \Phi(W),$$
$$H := SL_{n+1}(k) \cap \{\text{diagonal matrices}\}$$

Then $(G, (U_\alpha)_{\alpha \in \Phi}, H)$ is an RGD–system. (RGD0) is trivial, and (RGD1) follows from the commutator formulae for elementary matrices. The following three equalities imply (RGD2):

$$m\,U_\alpha m^{-1} = U_{s_a,\ell\alpha} \text{ for } m = \begin{pmatrix} 0 & ct^{-\ell} \\ -c^{-1}t^\ell & 0 \end{pmatrix}_{ij}, \quad c \in k^*, \ \ell \in \mathbb{Z} \text{ and } a = a_{ij}$$

$$\begin{pmatrix} 0 & \lambda \\ -\lambda^{-1} & 0 \end{pmatrix} = \begin{pmatrix} 1 & 0 \\ -\lambda^{-1} & 1 \end{pmatrix}\begin{pmatrix} 1 & \lambda \\ 0 & 1 \end{pmatrix}\begin{pmatrix} 1 & 0 \\ -\lambda^{-1} & 1 \end{pmatrix} \quad \text{and}$$

$$\begin{pmatrix} 0 & ct^{-\ell} \\ -c^{-1}t^\ell & 0 \end{pmatrix} = \begin{pmatrix} 0 & t^{-\ell} \\ -t^\ell & 0 \end{pmatrix}\begin{pmatrix} c^{-1} & 0 \\ 0 & c \end{pmatrix}$$

(RGD3) is a consequence of

$$HU_+ = \{(g_{rs}) \in G \mid g_{rs} \in k[t^{-1}] \text{ for } r \le s \text{ and } g_{rs} \in t^{-1}k[t^{-1}] \text{ for } r > s\},$$
$$U_- = \{(g_{rs}) \in G \mid g_{rs} \in tk[t] \text{ for } r < s, \ g_{rr} \in 1 + tk[t] \text{ and } g_{rs} \in k[t]$$
$$\text{for } r > s\}$$

Finally, (RGD4) is satisfied because G is generated by elementary matrices since $k[t, t^{-1}]$ is a Euclidean domain.

Example 3: Chevalley groups over $k[t, t^{-1}]$ (see also [Ab5], Lemma 5)

Keep the notations introduced before Example 2. Let $\mathcal{G}$ be a Chevalley group (scheme) of type Ψ, $\mathcal{T}$ a maximal torus of $\mathcal{G}$ and $\mathcal{N} := \mathcal{N}_\mathcal{G}(\mathcal{T})$. Identify the root system associated to $\mathcal{G}$ and $\mathcal{T}$ with Ψ . Denote by $\mathcal{U}_a$ the one dimensional unipotent subgroup of $\mathcal{G}$ corresponding to the root $a \in \Psi$. Set $\mathcal{U}_+ := \langle \mathcal{U}_a \mid a \in \Psi_+\rangle$ and $\mathcal{U}_- := \langle \mathcal{U}_a \mid a \in \Psi_-\rangle$. Then $\mathcal{B}_+ := \mathcal{T}\mathcal{U}_+$ and $\mathcal{B}_- := \mathcal{T}\mathcal{U}_-$ are opposite Borel subgroups of $\mathcal{G}$. Select isomorphisms $x_a : \text{Add} \xrightarrow{\sim} \mathcal{U}_a$ ($a \in \Psi$, Add = additive group), defined over $\mathbb{Z}$, such that the constants in Chevalley's commutator formulae are integers and such that there exist homomorphisms $\varphi_a : SL_2 \longrightarrow \mathcal{G}$ satisfying

$$\varphi_a\left(\begin{pmatrix} 1 & \lambda \\ 0 & 1 \end{pmatrix}\right) = x_a(\lambda) \text{ and } \varphi_a\left(\begin{pmatrix} 1 & 0 \\ \lambda & 1 \end{pmatrix}\right) = x_{-a}(\lambda) .$$

All this can be achieved for example by using the explicit constructions described in [St], §§3 and 5. As Steinberg, we shall use the notations $w_a(\lambda) := x_a(\lambda)x_{-a}(-\lambda^{-1})x_a(\lambda)$ and $h_a(\lambda) := w_a(\lambda)w_a(1)^{-1}$. Setting

$$
\begin{aligned}
G &:= \mathcal{G}(k[t,t^{-1}])^+ := \langle x_a(\lambda) \mid a \in \Psi, \lambda \in k[t,t^{-1}]\rangle \le \mathcal{G}(k[t,t^{-1}]) \\
U_\alpha &:= \{x_a(ct^{-\ell}) \mid c \in k\} \quad \text{for} \quad \alpha = \alpha_{a,\ell} \in \Phi = \Phi(W) \\
H &:= \langle h_a(c) \mid a \in \Psi, c \in k^*\rangle ,
\end{aligned}
$$

we obtain an RGD–system $(G, (U_\alpha)_{\alpha \in \Phi}, H)$ with Coxeter group $W = W_{\mathrm{aff}}(\Psi)$. In order to establish (RGD1), one uses Chevalley's commutator formulae. (RGD2) is a consequence of the equalities

$$w_a(ct^{-\ell})U_\alpha w_a(ct^{-\ell})^{-1} = U_{s_{a,\ell}\alpha} \text{ for } a \in \Psi, c \in k^*, \ell \in \mathbb{Z}, \alpha \in \Phi,$$

$$w_a(\lambda) = w_{-a}(-\lambda^{-1}) = x_{-a}(-\lambda^{-1})x_a(\lambda)x_{-a}(-\lambda^{-1}) \text{ and}$$

$$w_a(ct^{-\ell}) = w_a(t^{-\ell})h_a(c^{-1})$$

which are easily deduced from the usual relations in Chevalley groups (cf. [St], §3, p. 30, and §6). (RGD3) follows from

$$HU_+ \cap U_- \subseteq \mathcal{G}(k[t^{-1}]) \cap \mathcal{G}(k[t]) = \mathcal{G}(k),$$

$$HU_+ \cap \mathcal{G}(k) \subseteq \mathcal{B}_+(k) \text{ and } U_- \cap \mathcal{G}(k) \subseteq \mathcal{U}_-(k) .$$

(RGD4) is satisfied by the definition of G .

If $\mathcal{G}$ is **simply connected**, the groups $G, H, B_+ := HU_+, B_- := HU_-$ and $N := \langle H, m(u) \mid u \in U_{\alpha_i} \setminus \{1\}, i \in I \rangle$ can be described more directly. Firstly $G = \mathcal{G}(k[t, t^{-1}])$, $H = \mathcal{T}(k)$ and $N = \mathcal{N}(k[t, t^{-1}])$. Secondly, there exist group homomorphisms $\rho_\varepsilon : \mathcal{G}(k[t^{-\varepsilon}]) \longrightarrow \mathcal{G}(k)$ induced by reduction mod $t^{-\varepsilon}$ (where $t^+ := t$ and $t^- := t^{-1}$) , and one obtains $B_\varepsilon = \rho_\varepsilon^{-1}(\mathcal{B}_\varepsilon(k))$ for $\varepsilon \in \{+, -\}$.

I mention in passing that the affine BN-pair (B_ε, N) in G is obtained by intersecting an affine BN-pair in $\mathcal{G}(k(t))^+$ with G ($\mathcal{G}$ need not be simply connected here). This affine BN-pair in $\mathcal{G}(k(t))^+$ results from the valuation of the root datum $(\mathcal{U}_a(k(t))_{a \in \Psi}$ induced by the discrete valuation ω_ε of $k(t)$ determined by $\omega_\varepsilon(k^*) = \{0\}$ and $\omega_\varepsilon(t^{-\varepsilon}) = 1$ ($\varepsilon \in \{+, -\}$) as described in [BrT1], §6.5. The corresponding **Bruhat–Tits building** coincides with the affine building associated to (G, B_ε, N, S) because $k[t, t^{-1}]$ is dense in $k(t)$ relative to ω_ε .

Example 4: Almost simple k-groups over $k[t, t^{-1}]$ (cf. [T11], §3.2)

Let $\mathcal{G}$ be an almost simple group, defined and isotropic over k . Applying the theory of reductive groups over arbitrary ground fields (cf. [BoT]), it is again possible to construct an RGD-system in $\mathcal{G}(k[t, t^{-1}])^+$. The details are technically more involved than in Example 3, especially if the relative root system of $\mathcal{G}$ is not reduced, and will not be given here. Instead, I will only recall how the twin BN-pair in $G = \mathcal{G}(k[t, t^{-1}])$ looks like if $\mathcal{G}$ is simply connected. Let $\mathcal{T}$ be a maximal k-split torus of $\mathcal{G}$ and denote by $\mathcal{C}$ its centralizer and by $\mathcal{N}$ its normalizer in $\mathcal{G}$. Choose two opposite minimal parabolic k-subgroups $\mathcal{B}_+$ and $\mathcal{B}_-$ of $\mathcal{G}$ such that $\mathcal{B}_+ \cap \mathcal{B}_- = \mathcal{C}$. Consider again the homeomorphisms $\rho_\varepsilon : \mathcal{G}(k[t^{-\varepsilon}]) \longrightarrow \mathcal{G}(k)$ defined by reduction mod $t^{-\varepsilon}$ and set $B_\varepsilon := \rho_\varepsilon^{-1}(\mathcal{B}_\varepsilon(k))$ for $\varepsilon \in \{+, -\}$. If furthermore $N := \mathcal{N}(k[t, t^{-1}])$, $\quad H := \mathcal{C}(k)$ and S is an appropriate set of generators of $W := N/H$ (S is uniquely determined by each of the two BN-pairs (B_ε, N)), then (G, B_+, B_-, N, S) is a twin BN-pair.

For classical groups, the existence of an RGD-system in $\mathcal{G}(k[t, t^{-1}])^+$ can be established by applying the relations stated in [BrT1], §10, without referring to the general theory of reductive groups. In Chapter III the RGD-axioms will be verified explicitly for the non-split classical groups occurring in Theorem C.

Example 5: Kac–Moody groups over k   (cf. [T8] and [T11], §3.3)

In [T8], a group functor $\mathcal{G}_D$ is associated to every system

$D = (\Lambda, (\alpha_i)_{i \in I}, (h_i)_{i \in I})$ consisting of a finitely generated free $\mathbb{Z}$-module Λ, a family $(\alpha_i)_{i \in I}$ of elements of Λ and a family $(h_i)_{i \in I}$ of elements of $\Lambda^\vee$, the $\mathbb{Z}$-dual of Λ , provided that $A = (A_{ij})_{i,j \in I} := ((\alpha_j, h_i))_{i,j \in I}$ is a generalized Cartan matrix. If A is even a Cartan matrix, $\mathcal{G}_D$ is a reductive Chevalley–Demazure group scheme. In general, the restriction of the group functor $\mathcal{G}_D$ to fields yields the "minimal" Kac–Moody groups of type D . We recall that $\mathcal{G}_D$ is generated by certain subfunctors $\mathcal{T}$ and $\mathcal{U}_\alpha, \alpha \in \Phi$. Here, $\mathcal{T}$ denotes the split torus scheme defined by $\mathcal{T}(R) = \mathrm{Hom}(\Lambda, R^*)$ for every commutative ring R. $\Phi = \Phi(A)$ is the set of "real roots" associated to A. It is identified with the set of roots of $\Sigma(W, S)$ where (W, S) is the Coxeter system belonging to the generalized Cartan matrix A . The Coxeter matrix $M = (m_{ij})_{i,j \in I}$ of (W, S) is defined by $m_{ii} = 1$ and $m_{ij} = 2, 3, 4, 6$ or ∞ for $i \neq j$ according as $A_{ij} A_{ji} = 0, 1, 2, 3$ or $A_{ij} A_{ji} \geq 4$ $(i, j \in I)$. As in [T8], Section 3, $\mathcal{U}_\alpha$ denotes the group scheme associated to the root $\alpha \in \Phi$. It is isomorphic over $\mathbb{Z}$ to the additive group scheme Add.

Inserting k into the various group functors, one obtains the RGD-system $(\mathcal{G}_D(k), (\mathcal{U}_\alpha(k))_{\alpha \in \Phi}, \mathcal{T}(k))$. The RGD-axioms are verified in [T8]. (RGD0) and (RGD4) are obvious according to the definitions; (RGD1) and (RGD2) follow, similarly as in the case of Chevalley groups over k , from appropriate "Steinberg relations". But (RGD3) is more difficult here because it cannot be established by simply considering the shape of certain matrices. Instead, only the weaker condition "$U_{-\alpha_i} \not\subseteq U_+$ (and $U_{\alpha_i} \not\subseteq U_-$) for all $i \in I$" is checked directly, and from this, (RGD3) has to be

20

deduced by different methods (cf. Remark 2).

Note that $\mathcal{T}(k)\,\mathcal{G}(k[t,t^{-1}])^+$ is a minimal Kac–Moody group of affine type over k in the sense of Tits for every Chevalley group $\mathcal{G}$ with maximal torus $\mathcal{T}$. Hence the RGD-system discussed in Example 3 can be derived directly from the RGD-system above. However, I preferred including a more elementary treatment of $\mathcal{G}(k[t,t^{-1}])^+$ for readers interested in the latter group but not in the general theory of Kac–Moody groups.

§ 2 Twin buildings and twin apartments

To every Tits system (G, B, N, S), a thick building $\Delta = \Delta(G, B)$ is associated, the chambers of Δ being the left cosets of B in G (cf. [T1], Theorem 3.2.6). Therefore, to a twin BN-pair (G, B_+, B_-, N, S), there belongs a pair of buildings $\Delta_+ = \Delta(G, B_+)$, $\Delta_- = \Delta(G, B_-)$. But due to the conditions (TBN1) and (TBN2), (Δ_+, Δ_-) is endowed with an additional structure. The latter in particular provides a symmetric opposition relation between the chambers of Δ_+ and the chambers of Δ_-. However, in order to axiomatize the properties of this opposition relation efficiently, it is advantageous to introduce a "codistance" function between the chambers of Δ_+ and of Δ_-.

Before recalling the precise definition of a "twin building", I will fix some notions and notations. Throughout this book, a "building" is always understood to be a chamber complex of finite rank in the sense of [T1], not necessarily thick, equipped with a set of subcomplexes, called apartments, which are Coxeter complexes and satisfy the usual axioms (listed as (B3) and (B4) in [T1]). A building is said to be **of type M**, if each apartment of Δ is isomorphic to $\Sigma(W, S)$, where (W, S) is a Coxeter system of type M (see §1). If for one apartment Σ of Δ an isomorphism between Σ and $\Sigma(W, S)$ is chosen, the function type : $\Sigma(W, S) \longrightarrow 2^I$, $wW_J \longmapsto I \setminus J$, extends uniquely to a morphism of chamber complexes $\Delta \longrightarrow 2^I$, also denoted by "type". We fix a **numbering of** Δ, i.e. a type function as described above, for every building Δ. Morphisms between buildings of type M are always required to be type-preserving. If Δ is a building of type M and $a \in \Delta$, the cotype of a is defined by cotype $(a) := I \setminus \text{type}(a)$. a is said to be **of spherical cotype** or simply **spherical** if W_J is finite for $J = \text{cotype}(a)$. Two chambers $c, d \in \Delta$ are called **i-adjacent** $(i \in I)$ if $c \cap d$ is of cotype $\{i\}$. A gallery $(c_0, \ldots, c_m)$ is said to be **of type** $(i_1, \ldots, i_m)$ if

c_{j-1} and c_j are i_j-adjacent for all $1 \leq j \leq m$.

For every building Δ , we denote by $\mathcal{C}(\Delta)$ its set of chambers. If Δ is of type M, $\mathcal{C}(\Delta)$ is a chamber system of type M in the sense of [T4]. Furthermore, one obtains a well defined W-**valued distance function** $\delta : \mathcal{C}(\Delta) \times \mathcal{C}(\Delta) \longrightarrow W$ by associating to each pair (c, d) of chambers the element $s_{i_1} \ldots s_{i_m} \in W$, where $(i_1, \ldots, i_m)$ is the type of any minimal gallery with origin c and extremity d . A pair $(\mathcal{C}, \delta)$ consisting of a set $\mathcal{C}$ and a function $\delta : \mathcal{C} \times \mathcal{C} \longrightarrow W$ belongs to a building of type M if and only if it satisfies the following conditions (cf. [T11], §2.1):

(Bu1) $\delta(c, d) = 1 \iff c = d$

(Bu2) If $\delta(c, d) = w \in W$ and $\delta(d, e) = s \in S$, then $\delta(c, e) \in \{w, ws\}$.
 If additionally $\ell(ws) > \ell(w)$,then $\delta(d, e) = ws$.

(Bu3) Given $c, d \in \mathcal{C}$ and $s \in S$, there exists an $e \in \mathcal{C}$ satisfying
 $\delta(d, e) = s$ and $\delta(c, e) = ws$.

This characterization of buildings is, as far as I know, for the first time explicitly stated in [T9], but it is essentially already contained in [T4], Section 3. It motivates (together with Example 6 below) the definition of "twin buildings". The W-distance δ associated to a building of type M will repeatedly be used in the following. However, I do not adapt the point of view here that a building "is" a chamber system with certain additional properties. I rather stick to the "classical" definition of a building as given in [T1] because I think that simplicial complexes are closer to the geometric intuition than chamber systems, and this will be helpful in the following.

Definition 3 (cf. [T11], §2.2): *Let* Δ_+, Δ_- *be two buildings of type* M *with chamber sets* $C_+ = \mathcal{C}(\Delta_+)$, $C_- = \mathcal{C}(\Delta_-)$ *and* W-*distances* δ_+, δ_- . *Let furthermore a* W-**codistance** $\delta^* : C_+ \times C_- \cup C_- \times C_+ \longrightarrow W$ *be given. The triple* $(\Delta_+, \Delta_-, \delta^*)$ *is called a* **twin building of type** M *if the following conditions are satisfied for all* $c_\varepsilon \in C_\varepsilon$, $d_{-\varepsilon}, e_{-\varepsilon} \in C_{-\varepsilon}$ *and* $\varepsilon \in \{+, -\}$:

(Tw1) $\delta^*(d_{-\varepsilon}, c_\varepsilon) = \delta^*(c_\varepsilon, d_{-\varepsilon})^{-1}$

(Tw2) *If* $\delta^*(c_\varepsilon, d_{-\varepsilon}) = w$, $\delta_{-\varepsilon}(d_{-\varepsilon}, e_{-\varepsilon}) = s \in S$ *and* $\ell(ws) < \ell(w)$,
 then $\delta^*(c_\varepsilon, e_{-\varepsilon}) = ws$.

(Tw3) *If* $\delta^*(c_\varepsilon, d_{-\varepsilon}) = w$ *and* $s \in S$, *there exists an* $x_{-\varepsilon} \in C_{-\varepsilon}$ *such that* $\delta_{-\varepsilon}(d_{-\varepsilon}, x_{-\varepsilon}) = s$ *and* $\delta^*(c_\varepsilon, x_{-\varepsilon}) = ws$.

Two chambers $c_+ \in C_+$, $c_- \in C_-$ *are called* **opposite** *if* $\delta^*(c_+, c_-) = 1$; *this is also* denoted by "c_+ op c_-" or "c_- op c_+". Two simplices $a_+ \in \Delta_+$, $a_- \in \Delta_-$ are called opposite if they are of the same type and contained in opposite chambers.

If $\Sigma = \Sigma(W, S)$ is the Coxeter complex associated to (W, S), we obtain a twin building by setting $\Sigma_+ = \Sigma_- = \Sigma$ and

$$\delta^*(w_1, w_2) = \delta_+(w_1, w_2) = \delta_-(w_1, w_2) = w_1^{-1} w_2 \text{ for all } w_1, w_2 \in W = \mathcal{C}(\Sigma) .$$

Up to isomorphism, this is the only way how the structure of a twin building can be imposed on a pair of Coxeter complexes of type M (consider (Tw3)). Analogously to Example 1, every building of spherical type can also be interpreted as a twin building (cf. [T11], Proposition 1). We are mainly interested in twin buildings here because of the following

Example 6: The twin building associated to a twin BN-pair
(cf. [T11], §3.2)
Let (G, B_+, B_-, N, S) be a twin BN-pair with Coxeter system (W, S) of type M. Set $\Delta_\varepsilon = \Delta(G, B_\varepsilon)$ for $\varepsilon \in \{+, -\}$. Then the W-distance $\delta_\varepsilon : \mathcal{C}_\varepsilon \times \mathcal{C}_\varepsilon \longrightarrow W$ can be described as follows:

$$\delta_\varepsilon(gB_\varepsilon, hB_\varepsilon) = w \iff g^{-1}h \in B_\varepsilon w B_\varepsilon \quad (g, h \in G)$$

Using the Birkhoff decomposition introduced in Lemma 1, the W-codinstance δ^* is similarly defined by

$$\delta^*(gB_\varepsilon, hB_{-\varepsilon}) := \beta_\varepsilon^{-1}(B_\varepsilon g^{-1} h B_{-\varepsilon}) \quad \text{for} \quad g, h \in G .$$

Now the TBN-axioms immediately imply the Tw-axioms. Hence $(\Delta_+, \Delta_-, \delta^*)$ is a twin building of type M; it will be denoted by $\Delta(G, B_+, B_-)$ in the following. Note that two chambers $c_+ \in \Delta_+$ and $c_- \in \Delta_-$ are opposite if and only if there exists a $g \in G$ such that $c_+ = gB_+$ and $c_- = gB_-$.

In the rest of this section, $\Delta = (\Delta_+, \Delta_-, \delta^*)$ denotes a twin building of type M.

Remark 3: δ^* is already determined by Δ_+, Δ_- and the opposition relation because $\delta^*(c_\varepsilon, d_{-\varepsilon})$ is the unique element of minimal length in
$\{\delta_{-\varepsilon}(c_{-\varepsilon}, d_{-\varepsilon}) \mid c_{-\varepsilon} \in \mathcal{C}_{-\varepsilon} \text{ and } c_{-\varepsilon} \text{ op } c_\varepsilon\}$ for $c_\varepsilon \in \mathcal{C}_\varepsilon$ and $d_{-\varepsilon} \in \mathcal{C}_{-\varepsilon}$. This assertion is stated as Proposition 1 in [T9], §2.2, and can easily be deduced from the Tw-axioms

as follows:

First of all, (Tw2) and (Tw3) imply $\delta^*(c_\varepsilon, z_{-\varepsilon}) \in \{w, ws\}$, whenever $\delta^*(c_\varepsilon, y_{-\varepsilon}) = w \in W$ and $\delta_{-\varepsilon}(y_{-\varepsilon}, z_{-\varepsilon}) = s \in S$. Using this and considering minimal galleries between $c_{-\varepsilon}$ and $d_{-\varepsilon}$, we obtain $\ell(\delta^*(c_\varepsilon, d_{-\varepsilon})) \leq \ell(\delta_{-\varepsilon}(c_{-\varepsilon}, d_{-\varepsilon}))$ for any chamber $c_{-\varepsilon}$ op c_ε . Furthermore, if these two lengths are equal, then necessarily $\delta^*(c_\varepsilon, d_{-\varepsilon}) = \delta_{-\varepsilon}(c_{-\varepsilon}, d_{-\varepsilon})$. On the other hand, repeated application of (Tw3) yields at least one chamber $c^0_{-\varepsilon}$ satisfying $\delta^*(c_\varepsilon, c^0_{-\varepsilon}) = 1$ and $\delta_{-\varepsilon}(c^0_{-\varepsilon}, d_{-\varepsilon}) = \delta^*(c_\varepsilon, d_{-\varepsilon})$.

It is an important feature of every twin building that it is equipped with a distinguished system of "twin apartments".

Definition 4 (cf. [T11], §3.2, and [AR]): *If $c_+ \in C_+$ and $c_- \in C_-$ are opposite, we denote by $\Sigma(c_\varepsilon, c_{-\varepsilon})$ for $\varepsilon \in \{+, -\}$ the chamber subcomplex of Δ_ε having $\{d_\varepsilon \in C_\varepsilon \mid \delta^*(c_{-\varepsilon}, d_\varepsilon) = \delta_\varepsilon(c_\varepsilon, d_\varepsilon)\}$ as its set of chambers. The pair $\Sigma\{c_+, c_-\} := (\Sigma(c_+, c_-), \Sigma(c_-, c_+))$ is called a* **twin apartment** *of Δ. We set*

$$\mathcal{A}_\varepsilon := \{\Sigma(x_\varepsilon, x_{-\varepsilon}) \mid x_\varepsilon \in C_\varepsilon, \ x_{-\varepsilon} \in C_{-\varepsilon}, \ x_\varepsilon \text{ op } x_{-\varepsilon}\} \text{ for } \varepsilon \in \{+, -\} \text{ and}$$

$$\mathcal{A} := \{\Sigma\{x_+, x_-\} \mid x_+ \in C_+, \ x_- \in C_-, \ x_+ \text{ op } x_-\} \ .$$

The following observations concerning twin apartments, also motivating this term, are due to Tits (cf. [T9], §3.2, and [T11], §2.3). They are fundamental for an understanding of twin buildings and will therefore be provided with full proofs below.

Lemma 2: *ε denotes again the sign + or - .*

i) *$\mathcal{A}_\varepsilon$ is a system of apartments for Δ_ε , i.e. every element of $\mathcal{A}_\varepsilon$ is a Coxeter complex of type M and for any two $c_\varepsilon, d_\varepsilon \in C_\varepsilon$, there exists a $\Sigma_\varepsilon \in \mathcal{A}_\varepsilon$ containing c_ε and d_ε .*

ii) *Given $c_+ \in C_+$ and $d_- \in C_-$, there exists a $\Sigma = (\Sigma_+, \Sigma_-) \in \mathcal{A}$ such that $c_+ \in \Sigma_+$ and $d_- \in \Sigma_-$. Σ is uniquely determined if c_+ and d_- are opposite.*

iii) *If $\Sigma = (\Sigma_+, \Sigma_-) \in \mathcal{A}$, then $(\Sigma_+, \Sigma_-, \delta^*_{|\Sigma})$ is a twin building.*

iv) *For any $\Sigma_\varepsilon \in \mathcal{A}_\varepsilon$, there exists exactly one $\Sigma_{-\varepsilon} \in \mathcal{A}_{-\varepsilon}$ such that $(\Sigma_+, \Sigma_-) \in \mathcal{A}$.*

Proof:

i) Given opposite chambers $c_+ \in C_+, c_- \in C_-$, we first have to show that $\Sigma(c_+, c_-)$
is a Coxeter complex of type M. To this end, we prove the following claims:

(1) For any $w \in W$, there is exactly one $x_+(w) \in C_+$ satisfying

$$\delta^*(c_-, x_+(w)) = \delta_+(c_+, x_+(w)) = w$$

(Tw3) and (Bu2) imply that there exists at least one $x_+(w)$ with the required
properties. The uniqueness of $x_+(w)$ is shown by induction on $\ell(w)$. Assume
$w = w's$ with $s \in S$ and $\ell(w) = \ell(w') + 1$. Choose by (Bu3) $x'_+ \in C_+$ with
$\delta_+(c_+, x'_+) = w'$ and $\delta_+(x'_+, x_+(w)) = s$. Then $\delta^*(c_-, x'_+) = w'$ by (Tw2). Hence
$x'_+ = x_+(w')$ by induction hypothesis. So $x_+(w)$ satisfies $\delta_+(x_+(w'), x_+(w)) = s$
and $\delta^*(c_-, x_+(w)) = w's$. In view of (Tw2), there is only one chamber $x_+(w)$
with these properties.

(2) $\delta_+(x_+(ws), x_+(w)) = s$ for $w \in W$ and $s \in S$

This was proved under (1) in case $\ell(ws) < \ell(w)$. If $\ell(ws) > \ell(w)$, we just
apply this result to ws instead of w .

By (1) and (2), the map $W \longrightarrow C_+$, $w \longmapsto x_+(w)$, is an injective morphism of
chamber systems with image $C(\Sigma(c_+, c_-))$. Hence it induces a type preserving
isomorphism between $\Sigma(W, S)$ and $\Sigma(c_+, c_-)$. Of course an analogous statement
is true for $\Sigma(c_-, c_+)$.

Now let $c_+, d_+ \in C_+$ be given. We are looking for a chamber c_- op c_+ such
that $d_+ \in \Sigma(c_+, c_-)$, i.e. $\delta^*(c_-, d_+) = \delta_+(c_+, d_+)$. By (Tw3), we can find a
chamber c_- satisfying $\delta^*(d_+, c_-) = \delta_+(c_+, d_+)^{-1}$. Hence $\delta^*(c_-, d_+) = \delta_+(c_+, d_+)$
by (Tw1). But then, by choosing a minimal gallery between c_+ and d_+ and
applying (Tw2) repeatedly, one automatically obtains $\delta^*(c_-, c_+) = 1$.

ii) Given $c_+ \in C_+, d_- \in C_-$, we now choose a chamber $c_- \in C_-$ satisfying
$\delta_-(c_-, d_-) = \delta^*(c_+, d_-)$. As above, $\delta^*(c_+, c_-) = 1$ then holds automatically.
Hence $c_+ \in \Sigma(c_+, c_-)$ and $d_- \in \Sigma(c_-, c_+)$.

The second statement in ii) is a consequence of the following assertion:

(3) If c_+ and c_- as well as x_+ and x_- are opposite chambers such that
$$c_+ \in \Sigma(x_+, x_-) \text{ and } c_- \in \Sigma(x_-, x_+) , \text{ then } \Sigma(c_\varepsilon, c_{-\varepsilon}) = \Sigma(x_\varepsilon, x_{-\varepsilon})$$
for $\varepsilon \in \{+, -\}$.

Let c_+, c_-, x_+, x_- be given as in (3). Then $\delta^*(c_+, x_-) = \delta_+(c_+, x_+) =: w$ and $\delta^*(c_-, x_+) = \delta_-(c_-, x_-) =: v$. In view of Remark 3, we firstly obtain $\ell(v) \le \ell(w) \le \ell(v)$ and then $v = w$. We now prove (3) by induction on $\ell(w)$. We can suppose $\varepsilon = +$. Since $\Sigma(c_+, c_-)$ and $\Sigma(x_+, x_-)$ are apartments, it suffices to show $\Sigma(x_+, x_-) \subseteq \Sigma(c_+, c_-)$. The case $w = 1$ is clear. Next we assume $w = s \in S$. Let $y_+ \in \mathcal{C}(\Sigma(x_+, x_-))$ be given and set $u := \delta_+(y_+, x_+) = \delta^*(y_+, x_-)$. Since c_+, x_+, y_+ are chambers of the apartment $\Sigma(x_+, x_-)$, it follows $\delta_+(y_+, c_+) = us$. We have to show $y_+ \in \Sigma(c_+, c_-)$, i.e. $\delta^*(y_+, c_-) = us$. In case $\ell(us) < \ell(u)$, the latter follows from (Tw2). So let us assume $\ell(us) > \ell(u)$ and choose by (Tw3) $c'_- \in \mathcal{C}_-$ such that $\delta_-(x_-, c'_-) = s$ and $\delta^*(y_+, c'_-) = us$. Repeated application of (Tw2) (and (Tw1)) yields $\delta^*(c'_-, x_+) = s$. This means $c'_- \in \Sigma(x_-, x_+)$. From $c_- \in \Sigma(x_-, x_+), s = \delta_-(c_-, x_-) = \delta_-(c'_-, x_-)$ and i), we now deduce $c'_- = c_-$ and hence $\delta^*(y_+, c_-) = us$.

Finally, let $w \neq 1$ be arbitrary. We choose a representation $w = sw'$ with $s \in S$ and $\ell(w) = \ell(w') + 1$. Next we choose, for $\eta \in \{+, -\}$, $z_\eta \in \mathcal{C}(\Sigma(x_\eta, x_{-\eta}))$ with $\delta_\eta(c_\eta, z_\eta) = s$ and $\delta_\eta(z_\eta, x_\eta) = w'$. Then also $\delta^*(z_\eta, x_{-\eta}) = w'$ for $\eta \in \{+, -\}$ and hence $\delta^*(z_+, z_-) = 1$ by (Tw2). Now the induction hypothesis implies $\Sigma(z_+, z_-) = \Sigma(x_+, x_-)$ and $\Sigma(z_-, z_+) = \Sigma(x_-, x_+)$. In particular, $c_\eta \in \Sigma(z_\eta, z_{-\eta})$ for $\eta \in \{+, -\}$ and therefore
$$\Sigma(c_+, c_-) = \Sigma(z_+, z_-) = \Sigma(x_+, x_-) \text{ by the case } w = s \text{ treated above.}$$

iii) Let $c_+, c_-, \Sigma(c_+, c_-) = \Sigma_+$ and $x_+(w)$ be as under i). Define for $v \in W$ $x_-(v) \in \Sigma(c_-, c_+) = \Sigma_-$ analogously. In order to verify that $(\Sigma_+, \Sigma_-, \delta^*_{|\Sigma})$ is a twin building, it remains to show $\delta^*(x_-(v), x_+(w)) = v^{-1}w$. Note that $\delta^*(x_+(v), x_-(v)) = 1$ by (Tw2) and (Tw1). Hence we can apply (3) (roles of c and x reversed) which yields $\Sigma(x_+(v), x_-(v)) = \Sigma(c_+, c_-)$. Therefore $x_+(w) \in \Sigma(x_+(v), x_-(v))$ and hence $\delta^*(x_-(v), x_+(w)) = \delta_+(x_+(v), x_+(w)) = v^{-1}w$, the latter by (2).

iv) We can assume $\varepsilon = +$ and $\Sigma_+ = \Sigma(c_+, c_-)$ with $c_+ \text{ op } c_-$. If $(\Sigma_+, \Sigma_-) \in \mathcal{A}$, then there exist opposite chambers x_+ and x_- such that $\Sigma_\eta = \Sigma(x_\eta, x_{-\eta})$ for

$\eta \in \{+,-\}$. We have to show $\Sigma(x_-, x_+) = \Sigma(c_-, c_+)$. In view of (3), it suffices to verify $c_- \in \Sigma(x_-, x_+)$. Let $y_- \in \Sigma_-$ be opposite to c_+ and set $w := \delta_-(c_-, y_-)$. According to (3), $(\Sigma_+, \Sigma_-) = (\Sigma(c_+, y_-), \Sigma(y_-, c_+))$. Hence we can find a chamber $z_+ \in \Sigma_+$ satisfying $\delta_+(z_+, c_+) = \delta^*(z_+, y_-) = w$. Then (Tw2) implies $\delta^*(z_+, c_-) = 1$. Since $z_+ \in \Sigma(c_+, c_-)$, this is only possible if $z_+ = c_+$. Hence $w = 1$ and $c_- = y_- \in \Sigma_-$. $\square$

The first part of statement ii) above is analoguous to axiom (B3) of [T1]; the second generalizes a well known fact concerning spherical buildings (loc. cit., Proposition 3.25). In Lemma 3 (B4) will also be transferred to twin buildings. Before doing this, we mention an important consequence of Lemma 2 iii) and introduce isomorphisms of twin apartments.

Remark 4: If $\Sigma = (\Sigma_+, \Sigma_-)$ is a twin apartment, then for any chamber $c_\varepsilon \in \Sigma_\varepsilon$, there exists exactly one chamber $c_{-\varepsilon} \in \Sigma_{-\varepsilon}$ opposite to c_ε. The bijection between $\mathcal{C}(\Sigma_+)$ and $\mathcal{C}(\Sigma_-)$ obtained in this way extends to a type-preserving isomorphism $\mathrm{op}_\Sigma : \Sigma_+ \xrightarrow{\sim} \Sigma_-$ of Coxeter complexes.

By an **isomorphism of twin apartments** we understand a pair of (type-preserving) isomorphisms of Coxeter complexes preserving δ^*. Given
$\Sigma = (\Sigma_+, \Sigma_-)$, $\tilde{\Sigma} = (\tilde{\Sigma}_+, \widetilde{\Sigma_-}) \in \mathcal{A}$ and isomorphisms $\alpha_\varepsilon : \Sigma_\varepsilon \longrightarrow \tilde{\Sigma}_\varepsilon$ for $\varepsilon \in \{+, -\}$, the pair $\alpha = (\alpha_+, \alpha_-) : \Sigma \longrightarrow \tilde{\Sigma}$ is an isomorphism of twin apartments if and only if $\alpha_- \circ \mathrm{op}_\Sigma = \mathrm{op}_{\tilde{\Sigma}} \circ \alpha_+$ (see Remark 3).

Lemma 3: Given $\Sigma = (\Sigma_+, \Sigma_-), \tilde{\Sigma} = (\tilde{\Sigma}_+, \tilde{\Sigma}_-) \in \mathcal{A}$ and $a_+ \in \Sigma_+ \cap \tilde{\Sigma}_+$, $a_- \in \Sigma_- \cap \tilde{\Sigma}_-$, there exists an isomorphism of twin apartments $\alpha = (\alpha_+, \alpha_-) : \Sigma \longrightarrow \tilde{\Sigma}$ satisfying $\alpha_+(a_+) = a_+$ and $\alpha_-(a_-) = a_-$.

Proof: In view of Lemma 2 ii), it is sufficient to prove the claim under the additional assumption that a_+ or a_- is a chamber, say $a_+ \in \mathcal{C}_+$. According to ordinary building theory, there exists an isomorphism $\alpha_+ : \Sigma_+ \xrightarrow{\sim} \tilde{\Sigma}_+$ fixing a_+. In view of Remark 4, there is exactly one isomorphism $\alpha_- : \Sigma_- \xrightarrow{\sim} \tilde{\Sigma}_-$ such that $\alpha = (\alpha_+, \alpha_-) : \Sigma \longrightarrow \tilde{\Sigma}$ preserves δ^*, namely $\alpha_- = \mathrm{op}_{\tilde{\Sigma}} \circ \alpha_+ \circ \mathrm{op}_\Sigma^{-1}$. We have to show $\alpha_-(a_-) = a_-$. If a_- is a chamber, this follows from $\delta^*(a_+, \alpha_-(a_-)) = \delta^*(\alpha_+(a_+), \alpha_-(a_-)) = \delta^*(a_+, a_-)$, $a_+ \in \tilde{\Sigma}_+$, $a_-, \alpha_-(a_-) \in \tilde{\Sigma}_-$ and Lemma 2 iii).

Next, we assume that a_- is a panel, i.e. a simplex of codimension 1 in Δ_-. Let

$\{i\}$ be the cotype of a_- and $s := s_i \in S$. In view of (Tw2) and (Tw3),

$$\{\delta^*(a_+, x_-) \mid x_- \in \mathcal{C}_- \text{ and } a_- \subset x_-\} = \{w, ws\} \text{ for some } w \in W .$$

We may assume $\ell(ws) < \ell(w)$. Again by (Tw2), there is exactly one chamber $c_- \in \mathcal{C}_-$ such that $a_- \subset c_-$ and $\delta^*(a_+, c_-) = w$ (c_- is the "coprojection of a_+ onto a_-", see §4). Now the same reasoning applies to the twin buildings $(\Sigma_+, \Sigma_-, \delta^*_{|\Sigma})$ and $(\tilde{\Sigma}_+, \tilde{\Sigma}_-, \delta^*_{|\tilde{\Sigma}})$. Therefore, $c_- \in \Sigma_- \cap \tilde{\Sigma}_-$, implying $\alpha_-(c_-) = c_-$ as above and hence also $\alpha_-(a_-) = a_-$.

Finally, if a_- is an arbitrary simplex, we choose two chambers $x_- \in \Sigma_-$ and $\tilde{x}_- \in \tilde{\Sigma}_-$ containing a_- . Join x_- and $\tilde{x}_-$ by a gallery $x_- = x_0, x_1, \ldots, x_m = \tilde{x}_-$ in the star of a_- and choose twin apartments $\Sigma = \Sigma_0, \Sigma_1, \ldots, \Sigma_m = \tilde{\Sigma}$ such that $a_+ \in (\Sigma_j)_+$ and $x_j \in (\Sigma_j)_-$ for all $0 \leq j \leq m$. By the paragraph above, there exist isomorphisms $\alpha_j : \Sigma_{j-1} \xrightarrow{\sim} \Sigma_j$ fixing a_+ and $x_{j-1} \cap x_j$ ($1 \leq j \leq m$) . Hence $\alpha' := \alpha_m \circ \ldots \circ \alpha_1 : \Sigma \xrightarrow{\sim} \tilde{\Sigma}$ fixes a_+ and a_- . Note that $\alpha'_+(a_+) = \alpha_+(a_+)$ implies $\alpha'_+ = \alpha_+$, and this forces $\alpha'_- = \alpha_-$ as explained at the beginning of the proof. □

The statements of Lemma 2 ii), first sentence, and of Lemma 3 can be completed by further conditions in order to obtain a characterization of twin buildings by twin apartments (cf. [AR]), but this will not be used in the following.

We are now turning to group actions on twin buildings (cf. also [T11], §3.2). We say that a group G acts on $\Delta = (\Delta_+, \Delta_-, \delta^*)$ if it acts by (type-preserving) automorphisms on Δ_+ as well as on Δ_- , thereby preserving the codistance δ^* . Because of Remark 3, the latter is already satisfied if G preserves the opposition relation. In the theory of buildings, "strongly transitive" actions are particularly interesting (cf. [Br3], Ch. V, or [Ro], Ch. 5). The natural analogue of this notion in the case of twin buildings is the following:

Definition 5: *The action of a group G on a twin building Δ is called* **strongly transitive** *if the induced action on* $\{(c_+, c_-) \in \mathcal{C}_+ \times \mathcal{C}_- \mid c_+ \text{ op } c_-\}$ *is transitive.*

If (G, B_+, B_-, N, S) is a twin BN-pair, the action of G on $\Delta(G, B_+, B_-)$ is strongly transitive as is pointed out in Example 6. Note that the group B_ε is the stabilizer of the chamber $B_\varepsilon \in \Delta(G, B_\varepsilon)$ for $\varepsilon \in \{+, -\}$. The group N stabilizes the twin apartment $\Sigma = \Sigma\{B_+, B_-\}$ and acts transitively on $\mathcal{C}(\Sigma(B_\varepsilon, B_{-\varepsilon})) = \{nB_\varepsilon \mid n \in N\}$

for $\varepsilon \in \{+, -\}$. But N may be smaller than $\mathrm{Stab}_G(\Sigma)$. In fact, $\mathrm{Stab}_G(\Sigma) = N(B_+ \cap B_-)$. This may be deduced either directly from Definition 1 or from the second statement in Lemma 2 ii). The twin BN-pair is call **saturated** if $B_+ \cap B_- \subseteq N$. For example, every twin BN-pair belonging to an RGD-system is saturated.

If conversely a strongly transitive action of a group G on a thick twin building $(\Delta_+, \Delta_-, \delta^*)$ is given, a (saturated) twin BN-pair (G, B_+, B_-, N, S) can be constructed as follows (cf. [T11], §3.2, or [Ab5], Proposition 2): choose opposite chambers $c_+ \in \Delta_+$, $c_- \in \Delta_-$, and set
$B_\varepsilon := \mathrm{Stab}_G(c_\varepsilon)$ $(\varepsilon \in \{+, -\})$ and $N := \mathrm{Stab}_G(\Sigma\{c_+, c_-\})$.

We conclude this section by listing some simple properties of strongly transitive actions which will be applied later.

Lemma 4: *Assume that the group G acts strongly transitively on the twin building Δ. Then the following holds:*

i) *G acts transitively on $\mathcal{A}_+, \mathcal{A}_-$ and $\mathcal{A}$.*

ii) *For any twin apartment $\Sigma = (\Sigma_+, \Sigma_-)$,*
$\mathrm{Stab}_G(\Sigma_+) = \mathrm{Stab}_G(\Sigma) = \mathrm{Stab}_G(\Sigma_-)$ *acts transitively on $\mathcal{C}(\Sigma_+)$ and on $\mathcal{C}(\Sigma_-)$.*

iii) *Every isomorphism between two twin apartments of Δ is given by multiplication with an element of G.*

Proof: i) follows directly from the definitions. Lemma 2 iv) implies $\mathrm{Stab}_G(\Sigma_+) = \mathrm{Stab}_G(\Sigma) = \mathrm{Stab}_G(\Sigma_-)$. Given $c_\varepsilon, d_\varepsilon \in \mathcal{C}(\Sigma_\varepsilon)$, we denote by $c_{-\varepsilon}, d_{-\varepsilon}$ the respective opposites in $\mathcal{C}(\Sigma_{-\varepsilon})$. By definiton, there exists a $g \in G$ satisfying $gc_\varepsilon = d_\varepsilon$ and $gc_{-\varepsilon} = d_{-\varepsilon}$. By the second part of Lemma 2 ii), g is an element of $\mathrm{Stab}_G(\Sigma)$. Now statement iii) follows from i), ii) and the fact that every isomorphism of twin apartments is already uniquely determined by the image of a single chamber (see Remark 4). □

§ 3 Fundamental domains for strongly transitive actions on twin buildings

Throughout §3, $\Delta = (\Delta_+, \Delta_-, \delta^*)$ denotes a twin building of type M and G a group acting strongly transitively on Δ.

The results of this section are contained in the paper [Ab5], where they are treated in the more general context of "pre-twin buildings". The main purpose is to describe "fundamental domains" for the action of G on Δ and for the actions of certain subgroups of G on Δ_+ or Δ_- (see Propositions 2 and 3). The key argument is provided by Lemma 5 below which for its part is merely a translation of Lemma 3 into the language of group theory.

Lemma 5: Let $\Sigma = (\Sigma_+, \Sigma_-)$ be a twin apartment of Δ containing simplices $a_+ \in \Sigma_+$ and $a_- \in \Sigma_-$. Set $P_\varepsilon := \mathrm{Stab}_G(a_\varepsilon)$ for $\varepsilon \in \{+, -\}$, and let N be a subgroup of $\mathrm{Stab}_G(\Sigma)$ acting transitively on $\mathcal{C}(\Sigma_+)$ and on $\mathcal{C}(\Sigma_-)$. Then it follows

$$N \cap P_- P_+ = (N \cap P_-)(N \cap P_+)$$

Proof: Given $n = p_- p_+ \in N \cap P_- P_+$, we set $b_+ := p_- a_+ = n a_+ \in \Sigma_+$. Because of Lemma 3 and Lemma 4 iii), there exists a $g \in G$ mapping $p_- \Sigma$ onto Σ and fixing b_+ as well as a_-:

$$\Sigma \xrightarrow{\;p_-\;} p_- \Sigma \xrightarrow{\;g\;} \Sigma$$

$$(a_+, a_-) \longmapsto (b_+, a_-) \longmapsto (b_+, a_-)$$

In particular, $g \in P_-$ and $g p_- \in \mathrm{Stab}_G(\Sigma)$. In view of the assumption made on N , there exists an $h \in G$ fixing all elements of Σ such that $g p_- h \in N$. Hence it follows $n_1 := g p_- h \in N \cap P_-$ and $n_1 a_+ = g p_- a_+ = g b_+ = b_+ = n a_+$. This implies $n \in N \cap n_1 P_+ = n_1 (N \cap P_+) \subseteq (N \cap P_-)(N \cap P_+)$. $\qquad\Box$

If a group acts strongly transitively on a building Θ , a simplicial fundamental domain for this action is obviously given by $\bar{c} := \{a \in \Theta \mid a \subseteq c\}$ for any chamber $c \in \Theta$. In the case of twin buildings, things are not quite as trivial but still very well describable. The following proposition yields in particular a subcomplex F of $\Delta_+ \times \Delta_-$ satisfying $GF = \Delta_+ \times \Delta_-$ and $gF \cap hF \subseteq \partial(gF) \cap \partial(hF)$ for all $g \neq h \in G$. Hence F is a "fundamental domain" in the usual sense for the action of G on $\Delta_+ \times \Delta_-$ if this product is considered as a polysimplicial complex. With regard to Proposition

3, a precise description of the "identifications on the boundary of F" induced by G is desirable and can in fact easily be derived from Lemma 5.

Proposition 2: Let $\Sigma = (\Sigma_+, \Sigma_-)$ be a twin apartment of Δ, $c_- \in C(\Sigma_-)$ and $\bar{c}_- := \{a_- \in \Delta_- \mid a_- \subseteq c_-\}$. Assume that $N \leq \mathrm{Stab}_G(\Sigma)$ acts transitively on $C(\Sigma_+)$ and $C(\Sigma_-)$. Then it holds:

i) $G(\Sigma_+ \times \bar{c}_-) = \Delta_+ \times \Delta_-$

ii) Any two pairs $(a_+, a_-), (a'_+, a'_-) \in \Sigma_+ \times \bar{c}_-$ lie in the same G-orbit if and only if $a'_- = a_-$ and $a'_+ \in (N \cap \mathrm{Stab}_G(a_-))a_+$.

Proof:

i) Given $(b_+, b_-) \in \Delta_+ \times \Delta_-$, we choose a twin apartment $\tilde{\Sigma} = (\tilde{\Sigma}_+, \tilde{\Sigma}_-)$ such that $b_+ \in \tilde{\Sigma}_+$ and $b_- \in \tilde{\Sigma}_-$ (cf. Lemma 2 ii)). By Lemma 4, there exists a $g \in G$ such that $g\tilde{\Sigma} = \Sigma$ and $gb_- \subseteq c_-$. Hence $g(b_+, b_-) \in \Sigma_+ \times \bar{c}_-$.

ii) Because G acts type-preservingly, we only have to show the following: If $a_+ \in \Sigma_+$ and $a'_+ = p_- a_+ \in \Sigma_+$ for some $p_- \in P_- := \mathrm{Stab}_G(a_-)$, then $a'_+ \in (N \cap P_-)a_+$. N acting transitively on $C(\Sigma_+)$ and a_+, a'_+ being of the same type, there exists an $n \in N$ satisfying $na_+ = a'_+ = p_- a_+$. Hence $n \in N \cap p_- P_+ \subseteq N \cap P_- P_+$, where $P_+ := \mathrm{Stab}_G(a_+)$. Lemma 5 implies $n \in (N \cap P_-)(N \cap P_+)$ and therefore $a'_+ = na_+ \in (N \cap P_-)a_+$. $\quad\square$

Proposition 2 has an interesting consequence concerning the action of $\mathrm{Stab}_G(a_-)$ on Δ_+ . In the following, a "**simplicial fundamental domain**" or "**sfd**" for the action of a group H on a simplicial complex Θ is understood to be a subcomplex $\Theta' \subseteq \Theta$ such that for every $x \in \Theta$, there exists exactly one $x' \in \Theta'$ satisfying $x = hx'$ for at least one $h \in H$.

Proposition 3: Let $\Sigma = (\Sigma_+, \Sigma_-), c_-$ and N be given as in Proposition 2. Assume $a_- \subseteq c_-$ and set $P_- := \mathrm{Stab}_G(a_-)$. Then every sfd D_+ for the action of $N \cap P_-$ on Σ_+ is also an sfd for the action of P_- on Δ_+ .

Proof: Set $B_- := \mathrm{Stab}_G(c_-) \subseteq P_-$. Proposition 2 i) implies $B_-(\Sigma_+ \times \{c_-\}) = \Delta_+ \times \{c_-\}$ and hence $B_-\Sigma_+ = \Delta_+$. Therefore

$P_- D_+ \supseteq B_- (N \cap P_-) D_+ = B_- \Sigma_+ = \Delta_+$. If $a_+, a'_+ \in D_+$ lie in the same P_--orbit, then the same is true for (a_+, a_-) and (a'_+, a_-) . According to Proposition 2 ii), $a'_+ \in (N \cap P_-) a_+$, and hence $a'_+ = a_+$ because D_+ is an sfd for the action of $N \cap P_-$ on Σ_+ . $\qquad\qquad\qquad\qquad\qquad\qquad\qquad\qquad\qquad\qquad\qquad\qquad\qquad\qquad\qquad$ □

Recall that the apartments of Δ_+ and Δ_- are isomorphic to $\Sigma(W, S)$, where (W, S) is a Coxeter system of type M . In the following, we fix an isomorphism $\Sigma(W, S) \xrightarrow{\sim} \Sigma_+$ by sending the chamber 1 to the unique chamber c_+ of Σ_+ which is opposite to $c_- \in \Sigma_-$. Since N acts as the full group of type-preserving automorphisms on $\Sigma_+ \cong \Sigma(W, S)$, we obtain a surjective homomorphism $\nu : N \longrightarrow W$. If J is the cotype of $a_- \subseteq c_-$ and $P_- = \mathrm{Stab}_G(a_-)$, then $\nu(N \cap P_-) = W_J$ because an element of $N \subseteq \mathrm{Stab}_G(\Sigma)$ stabilizes a_- if and only if it stabilizes $\mathrm{op}_\Sigma^{-1}(a_-) \subseteq c_+$. In order to apply Proposition 3, we therefore have to determine an sfd for the action of W_J on $\Sigma(W, S)$. This merely requires a reinterpretation of certain well known facts concerning Coxeter groups (cf. [Bou2], Ch. IV, §1, Exercise 3). A detailed proof of the following lemma can be found in [Ab5], Section 2.

Lemma 6:   *Let (W, S) be a Coxeter system of type M .*
Set $W^J := \{w \in W \mid w \text{ is of minimal length in } W_J w\}$ for $J \subseteq I$.
Then $\Sigma^J := \{w W_K \mid w \in W^J \text{ and } K \subseteq I\}$ is an sfd for the action of W_J on $\Sigma(W, S)$. If α_i $(i \in I)$ denotes again the unique root containing 1 but not s_i, then $\Sigma^J = \bigcap\limits_{i \in J} \alpha_i$. $\qquad\qquad\qquad\qquad\qquad\qquad$ □

Remark 5:   It is easy to prove that every sfd for the action of W_J on $\Sigma(W, S)$ containing the chamber 1 must coincide with Σ^J .
Hence $\{v \Sigma^J \mid v \in W_J\}$ is exactly the set of all sfd's for the action of W_J on $\Sigma(W, S)$.

Using Example 6 and Lemma 6, Proposition 3 has the following application:

Corollary 1:   *Let (G, B_+, B_-, N, S) be a twin BN-pair with Coxeter system (W, S) of type M . Set $P_L^\varepsilon := B_\varepsilon W_L B_\varepsilon$ for $L \subseteq I$ and $\varepsilon \in \{+, -\}$. Then for any $J \subseteq I$, $\Sigma_+^J := \{w P_K^+ \mid w \in W^J \text{ and } K \subseteq I\}$ is an sfd for the action of P_J^- on $\Delta_+ = \Delta(G, B_+)$.* $\qquad\qquad$ □

This statement appears already in [T7], Section 15, where it is derived purely group theoretically without using twin buildings. Also the following specialization of

Corollary 1 is mentioned there, though somewhat implicitly.

Corollary 2:   *Given a Chevalley group $\mathcal{G}$ of type Ψ and a field k , we consider the twin BN-pair $(G = \mathcal{G}(k[t, t^{-1}])^+, B_+, B_-, N, S)$ associated to the RGD-system introduced in Example 3 of §1. Recall that $W = W_{\mathrm{aff}}(\Psi)$ and $I = \{i \in \mathbb{N}_0 \,|\, 0 \le i \le n\}$ in this case. Then for $J := I \setminus \{0\}$, $W_J = W(\Psi)$ is the linear Weyl group of Ψ , and $\Sigma^J = \bigcap_{i \in J} \alpha_i$ is a closed Weyl chamber or "quartier" in $\Sigma(W, S)$. Identifying this Coxeter complex with the "standard apartment" of the (Bruhat–Tits) building $\Delta_+ = \Delta(G, B_+)$, Corollary 1 implies that Σ^J is an sfd for the action of $P_J^- = \mathcal{G}(k[t])^+ (:= \langle x_a(\lambda) \,|\, a \in \Psi, \lambda \in k[t] \rangle)$ on Δ_+ .*   □

This result was first proved (in the case of simply connected Chevalley groups for $\mathcal{G}(k[t]) = \mathcal{G}(k[t])^+$) by Soulé (cf. [So]). His proof depends on certain calculations concerning SL_{n+1} and on the technically complicated §9 of [BrT1]. In contrast to Proposition 3 above, Soulé's approach is therefore geometrically not very transparent. Note also that Proposition 3, or more precisely Corollary 1, can immediately be applied to all other examples discussed in §1.

The key of the proof of Corollary 2 as described above consists in the simple fact that $\mathcal{G}(k[t])^+$ is the stabilizer in G of a vertex of $\Delta_- = \Delta(G, B_-)$, the "twin" of Δ_+ . This observation is quite generally useful when the action of $\mathcal{G}(k[t])^+$ on Δ_+ is studied as the following section will show.

§ 4 Coprojections in twin buildings

In ordinary building theory, "projections" play an important role. The existence of projections of chambers onto lower dimensional simplices for example is equivalent to the famous "gate property" (cf. [T1], 2.30.6 and 3.19.6) which is one of the characteristic features of buildings. In the framework of [T1], projections are introduced in connection with galleries and convex subcomplexes of buildings. But projections can also be characterized by means of the W-distance (cf. [DS]). If one wants to introduce the dual notion of "coprojection" into the theory of twin buildings (there are good reasons for doing this, see for example §5 below), only the second approach can be transferred because connecting galleries between chambers of Δ_+ and chambers of Δ_- do not exist.

The starting point for a general definition of coprojections in twin buildings is the following consequence of the axioms (Tw2) and (Tw3) which was already used in the proof of Lemma 3: If $(\Delta_+, \Delta_-, \delta^*)$ is a twin building, $c_{-\epsilon} \in \Delta_{-\epsilon}$ a chamber and $b_\epsilon \in \Delta_\epsilon$ a panel, then there exists exactly one chamber $d_\epsilon \in \Delta_\epsilon$ containing b_ϵ such that $\delta^*(c_{-\epsilon}, d_\epsilon)$ is of maximal length in $\{\delta^*(c_{-\epsilon}, x_\epsilon) \mid x_\epsilon \in \mathcal{C}_\epsilon \text{ and } b_\epsilon \subseteq x_\epsilon\}$. This statement can easily be generalized (cf. Lemma 10 i)) by replacing the requirement "$b_\epsilon \in \Delta_\epsilon$ is a panel" by "$b_\epsilon \in \Delta_\epsilon$ is a simplex of spherical cotype". We call d_ϵ the **coprojection of $c_{-\epsilon}$ onto b_ϵ** and denote it by $\text{proj}^*_{b_\epsilon} c_{-\epsilon}$.

As in ordinary building theory, we also want to "project" lower dimensional simplices, not only chambers. A definition of $\text{proj}^*_{b_\epsilon} a_{-\epsilon}$ which seems to be quite natural will be given in case $a_{-\epsilon}$ and b_ϵ are both of spherical cotype. Some "nice" properties of coprojections can be deduced then, see in particular Proposition 4 below. If only b_ϵ is spherical, one may still think of possible definitions of $\text{proj}^*_{b_\epsilon} a_{-\epsilon}$ (see Lemma 10 ii)), but essential parts of the present section are not applicable in this case. On the other hand, if b_ϵ is not spherical, the notion of a "coprojection of $a_{-\epsilon}$ onto b_ϵ" does not make sense even if $a_{-\epsilon}$ is a chamber.

Before discussing coprojections in detail, we have to collect some facts concerning Coxeter groups. Let (W, S) be again a Coxeter system with $S = \{s_i \mid i \in I\}$ and length function $\ell = \ell_S : W \longrightarrow \mathbb{N}_0$. For any subset $X \subseteq W$, we call an element minimal (respectively, maximal) in X if it is of minimal (respectively, maximal) length in X . The following statements are well known (cf. for example [Bou2], Ch. IV, §1, Exercises 3 and 22):

Lemma 7: *Assume $J, K \subseteq I$ and $w \in W$. Then it holds:*

i) *The minimal elements w_1 and w_2 in $W_J w$ and in $w W_K$ are uniquely determined. One obtains $\ell(w'w_1) = \ell(w') + \ell(w_1)$ for all $w' \in W_J$ and $\ell(w_2 w'') = \ell(w_2) + \ell(w'')$ for all $w'' \in W_K$.*

ii) *There exists a unique minimal element w_0 in $W_J w W_K$. If w is minimal in $W_J w$ and in $w W_K$, then $w_0 = w$.*

iii) *If W_J is finite, there exists a unique maximal element in W_J , called w_J^0 in the following. One has $w_J^0 \{s_j \mid j \in J\} w_J^0 = \{s_j \mid j \in J\}$ and $\ell(w' w_J^0) = \ell(w_J^0) - \ell(w') = \ell(w_J^0 w')$ for all $w' \in W_J$.* □

The next lemma can be derived from the previous one. However, I shall give a different, very short proof using Coxeter complexes.

Lemma 8: *Let w be minimal in $W_J w W_K$ and set*
$$J \cap wKw^{-1} := \{j \in J \mid s_j = w s_k w^{-1} \text{ for some } k \in K\} \ . \ \text{Then}$$

$$W_J \cap wW_Kw^{-1} = W_{J \cap wKw^{-1}} \ .$$

Proof: Interpret W_J and wW_K as elements of $\Sigma(W,S)$. Then $W_J \cap wW_Kw^{-1}$ is the stabilizer of the projection of wW_K onto W_J (cf. [T1], Proposition 12.5). On the other hand, this projection is the face of cotype $J \cap wKw^{-1}$ of the chamber 1 (cf. [DS], Proposition 3). □

Also the next result is probably not new. However, since everything else in this section depends on it and I do not know of an appropriate reference, it will be proved completely here.

Lemma 9: *Assume that W_J and W_K are finite. Suppose further that w is minimal in $W_J w W_K$ and set $J' := J \cap wKw^{-1}$, $K' := w^{-1}Jw \cap K$. Then $w^0 := w_J^0 w_{J'}^0 w \, w_K^0 = w_J^0 \, w \, w_{K'}^0 w_K^0$ is the unique maximal element in $W_J w W_K$.*

Proof: $w^{-1} w_{J'}^0 w = w_{K'}^0$ follows from $w^{-1}\{s_j \mid j \in J'\}w = \{s_k \mid k \in K'\}$. Now let x be an arbitrary maximal element in $W_J w W_K$. We shall show $x = w_J^0 w_{J'}^0 w w_K^0$. Because x is maximal in $W_J x$ as well as in $x W_K$, one obtains $w_J^0 x_1 = x = x_2 w_K^0$ for the minimal elements x_1 and x_2 in $W_J x$ and in $x W_K$ (cf. Lemma 7). Let z_1 be the minimal element in $x_1 W_K$ and set $y_1 := z_1^{-1} x_1 \in W_K$. Because of $\ell(x_1) = \ell(z_1) + \ell(y_1)$ and the minimality of x_1 , z_1 is also minimal in $W_J z_1$ and coincides therefore with w by Lemma 7 ii). Analogously, $x_2 = y_2 w$ with $y_2 \in W_J$. To sum up, $x = w_J^0 w y_1 = y_2 w w_K^0$ and $\ell(x) = \ell(w_J^0) + \ell(w) + \ell(y_1) = \ell(y_2) + \ell(w) + \ell(w_K^0)$. Furthermore, $y_2^{-1} w_J^0 = w w_K^0 y_1^{-1} w^{-1} \in W_J \cap wW_Kw^{-1} = W_{J'}$ by Lemma 8 and hence $y_2 \in w_J^0 W_{J'}$. The unique minimal element in $w_J^0 W_{J'}$ is $w_J^0 w_{J'}^0$ by Lemma 7 iii). Set $u := (w_J^0 w_{J'}^0)^{-1} y_2 \in W_{J'}$. The following comparison of lengths implies $u = 1$ and hence $x = w_J^0 w_{J'}^0 w w_K^0$:

$$\ell(w_J^0 w_{J'}^0) + \ell(u) + \ell(w) + \ell(w_K^0) = \ell(y_2) + \ell(w) + \ell(w_K^0) = \ell(x) =$$
$$= \ell(w_J^0 w_{J'}^0 u w w_K^0) = \ell(w_J^0 w_{J'}^0 w(w^{-1} u w w_K^0)) \leq \ell(w_J^0 w_{J'}^0) + \ell(w) + \ell(w_K^0) \ .$$

□

For the rest of this section, $\Delta = (\Delta_+, \Delta_-, \delta^*)$ denotes a twin building of type $M = (m_{ij})_{i,j \in I}$, $a_{-\epsilon} \in \Delta_{-\epsilon}$ a simplex of spherical cotype J and $b_\epsilon \in \Delta_\epsilon$ a simplex of spherical cotype K. We set

$$\mathcal{C}(a_{-\epsilon}) := \{c_{-\epsilon} \in \mathcal{C}_{-\epsilon} \mid a_{-\epsilon} \subseteq c_{-\epsilon}\} \text{ and } \mathcal{C}(b_\epsilon) := \{d_\epsilon \in \mathcal{C}_\epsilon \mid b_\epsilon \subseteq d_\epsilon\} .$$

The Tw-axioms immediately imply $\delta^*(\mathcal{C}(a_{-\epsilon}) \times \mathcal{C}(b_\epsilon)) = W_J w W_K$ for some $w \in W$. In the following, we may and shall assume that w is minimal in $W_J w W_K$. The maximal element in $W_J w W_K$, uniquely determined by Lemma 9, will again be denoted by w^0. Finally, we set

$$\mathcal{C}(b_\epsilon; a_{-\epsilon}) := \{d_\epsilon \in \mathcal{C}(b_\epsilon) \mid \exists\, c_{-\epsilon} \in \mathcal{C}(a_{-\epsilon}) \text{ such that } \delta^*(c_{-\epsilon}, d_\epsilon) = w^0\}.$$

Definition 6:   *The **coprojection** of $a_{-\epsilon}$ onto b_ϵ is the simplex*

$$\text{proj}^*_{b_\epsilon}\, a_{-\epsilon} := \bigcap_{d_\epsilon \in \mathcal{C}(b_\epsilon; a_{-\epsilon})} d_\epsilon \in \Delta_\epsilon .$$

Elementary facts concerning coprojections, very similar to the properties of projections discussed in [DS], are listed in the lemma below. Without loss of generality, we shall assume $\epsilon = +$ from now on.

Lemma 10:

i) *If a_- is a chamber, then $\text{proj}^*_{b_+}\, a_-$ is a chamber, too.*

ii) *$\{\text{proj}^*_{b_+}\, c_- \mid c_- \in \mathcal{C}(a_-)\} = \mathcal{C}(b_+; a_-)$. In particular*
$$\text{proj}^*_{b_+}\, a_- = \bigcap_{c_- \in \mathcal{C}(a_-)} \text{proj}^*_{b_+}\, c_- .$$

iii) *Assume $d_+ \in \mathcal{C}(b_+; a_-)$ and $d'_+ \in \mathcal{C}(b_+)$. Then $d'_+ \in \mathcal{C}(b_+; a_-)$ if and only if $\delta_+(d_+, d'_+) \in W_{K''}$, where $K'' := w^0_K(w^{-1}Jw \cap K)w^0_K \subseteq K$.*
Hence $\mathcal{C}(b_+; a_-) = \mathcal{C}(\text{proj}^*_{b_+}\, a_-)$ and cotype $(\text{proj}^*_{b_+}\, a_-) = K''$.

Proof:

i) Choose a $d_+ \in \mathcal{C}(b_+; a_-)$, i.e. $\delta^*(a_-, d_+) = w^0 = w w^0_K$. Let d'_+ be an arbitrary element of $\mathcal{C}(b_+)$. Then $v := \delta_+(d_+, d'_+) \in W_K$, and hence $\ell(w^0 v) = \ell(w^0) - \ell(v)$ by Lemma 7. Therefore (Tw2) implies $\delta^*(a_-, d'_+) = w^0 v$. In particular, $\mathcal{C}(b_+; a_-) = \{d_+\}$ and $\text{proj}^*_{b_+}\, a_- = d_+$.

36

ii) If $d_+ \in \mathcal{C}(b_+; a_-)$, there exists a $c_- \in \mathcal{C}(a_-)$ such that $\delta^*(c_-, d_+) = w^0$. In particular, $\delta^*(c_-, d_+)$ is maximal in $\delta^*(\{c_-\} \times \mathcal{C}(b_+))$, and hence $d_+ = \mathrm{proj}_{b_+} c_-$.

Conversely, let $c_- \in \mathcal{C}(a_-)$ be given, and set $d_+ := \mathrm{proj}_{b_+}^* c_-$. We have to show $w^0 \in \delta^*(\mathcal{C}(a_-) \times \{d_+\})$. Recall that $\delta^*(\{c_-\} \times \mathcal{C}(b_+)) = uwW_K$ for some $u \in W_J$. We choose u minimal with this property. Then uw is minimal in uwW_K , because otherwise $\ell(uws_k) < \ell(uw)$ for at least one $k \in K$. But w being minimal in wW_K , this would imply $uws_k = u'w$ with $u' \in W_J$ and $\ell(u') < \ell(u)$, contradicting the choice of u . Now the minimality of uw in uwW_K implies the maximality of uww_K^0 in $uwW_K = \delta^*(\{c_-\} \times \mathcal{C}(b_+))$. Hence $uww_K^0 = \delta^*(c_-, \mathrm{proj}_{b_+}^* c_-) = \delta^*(c_-, d_+)$ and $w^0 = (w_J^0 w_{J'}^0 u^{-1}) uww_K^0 \in \delta^*(\mathcal{C}(a_-) \times \{d_+\})$.

iii) $d_+ \in \mathcal{C}(b_+; a_-)$ implies $\delta^*(\mathcal{C}(a_-) \times \{d_+\}) = W_J w^0 = W_J ww_K^0$. Choose a $c_- \in \mathcal{C}(a_-)$ satisfying $\delta^*(c_-, d_+) = ww_K^0$. Setting $v := \delta_+(d_+, d'_+) \in W_K$, the equality $\delta^*(c_-, d'_+) = ww_K^0 v$ follows as under i). Now we obtain the following equivalences:

$$
\begin{aligned}
d'_+ \in \mathcal{C}(b_+; a_-) &\iff w^0 \in \delta^*(\mathcal{C}(a_-) \times \{d'_+\}) = W_J ww_K^0 v \\
&\iff ww_K^0 \in W_J ww_K^0 v \\
&\iff w_K^0 v w_K^0 \in W_K \cap w^{-1} W_J w \\
&\iff w_K^0 v w_K^0 \in W_{K \cap w^{-1} Jw} =: W_{K'} \text{ (by Lemma 8)} \\
&\iff v \in w_K^0 W_{K'} w_K^0 = W_{K''}
\end{aligned}
$$

(Recall that $w_K^0 K w_K^0 = K$ by Lemma 7 iii).) This proves the first claim, and the second follows immediately. $\qquad\square$

Corollary 3: Let $\Sigma = (\Sigma_+, \Sigma_-)$ be a twin apartment of Δ such that $b_+ \in \Sigma_+$ and $a_- \in \Sigma_-$. Recall that $(\Sigma_+, \Sigma_-, \delta^*_{|\Sigma})$ is a twin building as well (cf. Lemma 2 iii)). Then the coprojection $_\Sigma \mathrm{proj}_{b_+}^* a_-$ relative to Σ coincides with $\mathrm{proj}_{b_+}^* a_- = {}_\Delta \mathrm{proj}_{b_+}^* a_-$.

Proof: Since $\delta^*((\mathcal{C}(a_-) \cap \Sigma_-) \times (\mathcal{C}(b_+) \cap \Sigma_+)) = W_J w W_K$, $\mathcal{C}(b_+; a_-) \cap \Sigma_+$ is not empty and contains a chamber d_+ . Now by Lemma 10 iii), $_\Sigma \mathrm{proj}_{b_+}^* a_-$ as well as $_\Delta \mathrm{proj}_{b_+}^* a_-$ is the face of cotype K'' of d_+ . $\qquad\square$

Proposition 4: *Assume that $\Sigma = (\Sigma_+, \Sigma_-)$ is a twin apartment of Δ containing b_+ and a_-. Let op_Σ be as in Remark 4. Denote by "proj" the usual projection in buildings and by $\mathrm{op}_{\Sigma_{b_+}}$ the opposition involution of the spherical Coxeter complex $\Sigma_{b_+} := \{f_+ \in \Sigma_+ \mid b_+ \subseteq f_+\}$. Then*

$$\mathrm{proj}^*_{b_+} a_- = \mathrm{op}_{\Sigma_{b_+}} \left(\mathrm{proj}_{b_+} \left(\mathrm{op}_\Sigma^{-1}(a_-) \right) \right)$$

Proof: In view of the above corollary, we can assume $\Delta = \Sigma$.

Set $a_+ := \mathrm{op}_\Sigma^{-1}(a_-)$, and note that $\delta^*(\mathcal{C}(a_-) \times \{x_+\}) = \delta_+(\mathcal{C}(a_+) \times \{x_+\})$ for any $x_+ \in \mathcal{C}(\Sigma_+)$. On the other hand, $\delta_+(c_+, x_+)\delta_+(x_+, y_+) = \delta_+(c_+, y_+)$ for any three chambers c_+, x_+, y_+ of the Coxeter complex Σ_+. Therefore, for any $d_+ \in \mathcal{C}(b_+)$, $w w_K^0 \in \delta^*(\mathcal{C}(a_-) \times \{d_+\})$ if and only if $w \in \delta_+(\mathcal{C}(a_+) \times \{\mathrm{op}_{\Sigma_{b_+}}(d_+)\})$. This implies

$$\mathrm{op}_{\Sigma_{b_+}} (\mathcal{C}(b_+; a_-)) = \{d'_+ \in \mathcal{C}(b_+) \mid \exists\, c_+ \in \mathcal{C}(a_+) \text{ such that } \delta_+(c_+, d'_+) = w\}.$$

Since w is minimal in $W_J w W_K$, the set of chambers on the right hand is precisely $\mathcal{C}(\mathrm{proj}_{b_+} a_+)$ (cf. [DS], Section 3).

Hence $\mathrm{op}_{\Sigma_{b_+}} (\mathcal{C}(\mathrm{proj}^*_{b_+} a_-)) = \mathcal{C}(\mathrm{proj}_{b_+} a_+)$, and the claim follows. $\square$

I mention in passing the following consequence of Proposition 4:

The term $\mathrm{op}_{\Sigma_{b_+}} (\mathrm{proj}_{b_+}(\mathrm{op}_\Sigma^{-1}(a_-)))$ is independent of the choice of the twin apartment Σ containing b_+ and a_-. An application of Proposition 4 to groups acting strongly transitively on Δ will be given in §5, see Lemma 12 below. Here, I only demonstrate by means of an example that coprojections can be useful in order to study the geometric properties of group actions on certain affine buildings.

Example 7: Keep the notations introduced in §1, Example 3, and in §3, Corollary 2. Let $\Delta = (\Delta_+, \Delta_-, \delta^*)$ be the twin building associated to the twin BN-pair $(G = \mathcal{G}(k[t, t^{-1}])^+, B_+, B_-, N, S)$, and consider the action of $\Gamma := \mathcal{G}(k[t])^+$ on Δ_+. Then for any $b_+ \in \Delta_+$, $\Gamma_{b_+} := \mathrm{Stab}_\Gamma(b_+)$ automatically stabilizes a simplex which is in general strictly bigger than b_+. This fact can be explained very naturally by using coprojections. Indeed, recalling the identity $\Gamma = P^-_{I \setminus \{0\}} (= B_- W_{I \setminus \{0\}} B_-)$ and denoting the corresponding vertex of Δ_- by 0_-, it is clear in view of Definition 6 that $\Gamma_{b_+} = G_{0_-} \cap G_{b_+}$ stabilizes $\mathrm{proj}^*_{b_+} 0_-$ as well.

If b_+ is of the form σ_x as described in [Ab3], $\mathrm{proj}^*_{b_+} 0_-$ is nothing else but the simplex denoted by $\sigma_{[x,y]}$ there. This can be shown by combining Proposition 4 above with Lemma 3 in [Ab3]. The fact that Γ_{σ_x} stabilizes $\sigma_{[x,y]}$ can also be deduced from the following observation due to Soulé (cf. [So]): Let Σ_+ be the "standard apartment" of Δ_+ and identify its geometric realization $|\Sigma_+|$ with $\mathbb{R}^n$. Then for any $x \in \mathbb{R}^n \setminus \{0\}$, $\Gamma_x := \text{Stab}_\Gamma(x)$ fixes the whole ray $[x[:= \{\lambda x \mid \lambda \in \mathbb{R} \text{ and } \lambda \geq 1\}$ pointwise. Also this stronger statement admits a natural interpretation in the context of twin buildings, as we shall see after having introduced the notion of "coconvexity" below.

Appendix: Coconvexity

As already mentioned at the beginning of §4, projections and convex subcomplexes are closely related to each other in ordinary building theory. Having introduced coprojections, it is natural to look for an appropriate notion of "coconvexity" in twin buildings as well. Some results in this direction are presented in the following. The proofs are omitted here because this appendix will be referred to only once (cf. §6, Proposition 6) throughout the present book. They will be published elswhere.

Definition 7: *Let Θ_ϵ be a subcomplex of Δ_ϵ for $\epsilon \in \{+, -\}$. The pair (Θ_+, Θ_-) is called* **coconvex** *if* $\text{proj}_{b_\epsilon}^* a_{-\epsilon} \in \Theta_\epsilon$ *for any two spherical simplices* $b_\epsilon \in \Theta_\epsilon$ *and* $a_{-\epsilon} \in \Theta_{-\epsilon}$.

This definition is motivated by the following facts:

1. A chamber subcomplex κ of a building is convex if and only if
   $\text{proj}_b c \in \kappa$ for any panel $b \in \kappa$ and any chamber $c \in \kappa$.

2. An arbitrary subcomplex κ of a Coxeter complex is convex if and only if
   $\text{proj}_b a \in \kappa$ for any two simplices $a, b \in \kappa$.

Nevertheless, it is not quite obvious which definition of coconvexity is the "right" one. Definition 7 states the minimal condition every coconvex pair should satisfy. But this condition is "weak" if Θ_+ and Θ_- contain "few" spherical simplices. Furthermore, it is not clear even in the case of chamber subcomplexes whether the coconvexity of (Θ_+, Θ_-) implies that Θ_+ and Θ_- are convex. In fact, I have reason (though yet no counter-example) for doubting that this is true.

However, an important special case will be mentioned below where all these difficulties do not occur. Before, two further notions have to be explained:

1. Because $(\bigcap_j \Theta_+^j, \bigcap_j \Theta_-^j)$ is coconvex if all (Θ_+^j, Θ_-^j) are coconvex, every pair (Θ_+, Θ_-) of subsets of (Δ_+, Δ_-) is contained in a unique minimal coconvex pair of subcomplexes, called the **coconvex hull** of (Θ_+, Θ_-).

2. A pair of subcomplexes $\alpha = (\alpha_+, \alpha_-)$ is called a **twin root of** Δ if there exists a twin apartment $\Sigma = (\Sigma_+, \Sigma_-)$ such that α_ϵ is a root of Σ_ϵ for $\epsilon \in \{+, -\}$ and $\mathrm{op}_\Sigma(\alpha_+) = -\alpha_-$, the root opposite to α_- in Σ_- . Using Proposition 4, it is easy to check that every twin root is coconvex.

Proposition 5:   Let $b_+ \in \Delta_+$, $a_- \in \Delta_-$ be simplices of spherical cotype and $\Sigma = (\Sigma_+, \Sigma_-)$ a twin apartment of Δ containing b_+ and a_- . Then the coconvex hull of $(\{b_+\}, \{a_-\})$ is the intersection of all twin roots of Σ containing b_+ and a_- .

Proposition 5 goes well together with the characterization of convex subcomplexes of Coxeter complexes (cf. [T1], Theorem 2.19) and of full convex hulls of subsets of a building which are contained in an apartment (cf. [T1], Proposition 3.18). It is somewhat surprising that one need not presuppose the existence of any spherical simplices different from b_+ and a_- .

Corollary 4:   Let (Θ_+, Θ_-) be the coconvex hull of $(\{b_+\}, \{a_-\})$, and set $b_- := \mathrm{op}_\Sigma(b_+)$, $a_+ := \mathrm{op}_\Sigma^{-1}(a_+)$. Then it follows

$$\Theta_+ = \bigcap\{\alpha_+ \,|\, \alpha_+ \text{ is a root of } \Sigma_+, \, b_+ \in \alpha_+ \text{ and } a_+ \in -\alpha_+\}$$
$$\Theta_- = \bigcap\{\alpha_- \,|\, \alpha_- \text{ is a root of } \Sigma_-, \, b_- \in -\alpha_- \text{ and } a_- \in \alpha_-\}$$

In particular, Θ_+ and Θ_- are convex subcomplexes of rank

$$r := \mathrm{rk}\ \mathrm{proj}_{b_+}^* \, a_- = \mathrm{rk}\ \mathrm{proj}_{b_+} a_+ = \mathrm{rk}\ \mathrm{proj}_{a_-} b_- = \mathrm{rk}\ \mathrm{proj}_{a_-}^* \, b_+ \ .$$

Using some results about convex subcomplexes of Coxeter complexes (cf. [Ab2], Proposition 1), we obtain the following statement: Θ_ϵ is a chamber complex with the property that every simplex of rank $r - 1$ is contained in at most two simplices of rank r (Θ_ϵ is "thin with boundary"). This immediately implies

Corollary 5:   Let G be a group acting strongly transitively on Δ , and let (Θ_+, Θ_-) be given as in Corollary 4. Then $G_{b_+} \cap G_{a_-}$ fixes Θ_+ and Θ_- pointwise.

Example 7 (continued): We just specialize the last two corollaries to the following situation: a_- is the vertex 0_- of Δ_-, $a_+ = 0_+$ the vertex of Σ_+ opposite to 0_- and $b_+ \neq 0_+$ an arbitrary simplex of Σ_+. Identifying again $|\Sigma_+|$ with $\mathbb{R}^n$, 0_+ coincides with $0 \in \mathbb{R}^n$. Let $x \in \mathbb{R}^n$ be an element of the cell corresponding to b_+. Then by the Corollaries 4 and 5,

$\Gamma_x = \Gamma_{b_+} = G_{0_-} \cap G_{b_+}$ fixes all elements of

$$|\Theta_+| = \cap\{\alpha_{a,\ell} \,|\, a \in \Psi,\ \ell \in \mathbb{Z},\ (a,x) + \ell \geq 0 \text{ and } \ell \leq 0\}\ ,$$

where $\alpha_{a,\ell} = \{v \in \mathbb{R}^n \,|\, (a,v) + \ell \geq 0\}$ as in §1. In particular, $|\Theta_+|$ is a closed convex subset of $\mathbb{R}^n$ containing the ray $[x[$.

§ 5   A G_{a_-}-invariant filtration of Δ_+

We are now going to apply the results of the last two sections in order to deduce finiteness properties of $G_{a_-} = \mathrm{Stab}_G(a_-)$, where G is a group acting strongly transitively on a twin building $(\Delta_+, \Delta_-, \delta^*)$ subject to certain conditions (see §6). FP_m-and F_m-properties of a group are usually derived by studying the action of this group on an appropriate space. In the case of G_{a_-}, this will be the geometric realization $|\Delta_+|$ of Δ_+. Since the action of G_{a_-} on Δ_+ admits an sfd by Proposition 3, we can apply the concept of "Γ-restrictions" introduced by Abels in [A2] and also used in [Ab3]. We recall some notations and results in this context:

Let Δ be a building and $\mathcal{C}$ its set of chambers. Denote by $d : \mathcal{C} \times \mathcal{C} \longrightarrow \mathbb{N}_0$ the usual gallery-distance (i.e. $d = \ell \circ \delta : \mathcal{C} \times \mathcal{C} \longrightarrow W \longrightarrow \mathbb{N}_0$ if (W, S) is the Coxeter system associated to Δ). Set

$$d(a,b) := \min\{d(x,y) \,|\, x,y \in \mathcal{C},\ a \subseteq x \text{ and } b \subseteq y\} \quad \text{for } a, b \in \Delta \text{ and}$$
$$d(A,B) := \min\{d(a,b) \,|\, a \in A \text{ and } b \in B\} \qquad \text{for } A, B \subseteq \Delta\ .$$

Let Γ be a group acting on Δ, and suppose that there exists an sfd F for this action (this is the essential assumption in this context!). Denote by $r : \Delta \longrightarrow F$ the simplicial retraction mapping each $a \in \Delta$ onto the unique element of $\Gamma a \cap F$. Fix a chamber $c_0 \in F$, and set

$$
\begin{aligned}
F_j &:= \{a \in F \,|\, d(a, c_0) \leq j\}, \\
\Delta_j &:= \{b \in \Delta \,|\, d(b, \Gamma c_0) \leq j\} = \Gamma F_j \quad (j \in \mathbb{N}_0)
\end{aligned}
$$

We set $\bar{x} := \{y \in \Delta \mid y \subseteq x\}$ for any $x \in \Delta$. If c is a chamber contained in $\Delta_{j+1} \setminus \Delta_j$, then by [A2], Lemma 2.4, $\bar{c} \cap (\Delta_{j+1} \setminus \Delta_j)$ possesses a unique minimal element, denoted by $R^\Gamma(c)$ and called the Γ-restriction of c. We set

$$
\begin{aligned}
R_{j+1} &:= \{R^\Gamma(c) \mid c \in \mathcal{C} \text{ and } c \in \Delta_{j+1} \setminus \Delta_j\} \text{ and for any } b \in R_{j+1} \\
S(b) &:= \{a \in \Delta \mid a \cup b \in \Delta_{j+1}\} = st_{\Delta_{j+1}}(b) , \\
T'(b) &:= \{a \in S(b) \mid a \cap b = \emptyset\} = \ell k_{\Delta_{j+1}}(b) , \\
T(b) &:= S(b) \cap \Delta_j .
\end{aligned}
$$

In order to analyse the Γ-invariant filtration $(\Delta_j)_{j \in \mathbb{N}_0}$ of Δ, we need the four statements listed below. The first three are proved in [A2], Sections 2 and 4, and the last one in [Ab3], §2.2.

Lemma 11: *With the notations introduced above the following holds:*

i) $\Delta_{j+1} = \Delta_j \cup \bigcup\limits_{b \in R_{j+1}} S(b)$

ii) $S(b) \cap S(b') \subseteq \Delta_j$ *for all* $b \neq b' \in R_{j+1}$

iii) $T(b) = \partial b * T'(b)$ *, where* $\partial b := \bar{b} \setminus \{b\}$ *and "* $*$ *" means "join"*

iv) $T'(\gamma b) = \gamma T'(b)$ *for all* $\gamma \in \Gamma, b \in R_{j+1}$ *and*
  $T'(b) = \Gamma_b\{a \in \Delta \mid a \subseteq \mathrm{proj}_b \, c_0 \text{ and } a \cap b = \emptyset\}$ *for all* $b \in R_{j+1} \cap F$. $\square$

Now we consider the following specialization: Given a group G acting strongly transitively on a twin building $(\Delta_+, \Delta_-, \delta^*)$ of type M and an element $\emptyset \neq a_- \in \Delta_-$, we set $\Gamma = G_{a_-}$ and $\Delta = \Delta_+$. We choose a twin apartment $\Sigma = (\Sigma_+, \Sigma_-)$ with $a_- \in \Sigma_-$ and an sfd $F = D_+ \subseteq \Sigma_+$ for the action of G_{a_-} on Δ_+ as described in Proposition 3 (see also Lemma 6). Finally, c_0 is the (unique) chamber of D_+ containing $a_+ := op_\Sigma^{-1}(a_-)$.

In order to derive finiteness properties of G_{a_-} from the set-up described above, one has to analyse the homotopy properties of the various $|T'(b_+)|$. In view of the intended applications, we restrict ourselves to considering spherical simplices a_- and b_+. In this case, $T'(b_+)$ can be described more explicitly than in Lemma 11 iv), i.e. without referring to Γ_{b_+}. This is due to the following consequence of Proposition 4:

Lemma 12: *Assume that a_- and $b_+ \in D_+$ are of spherical cotype. Set $p_+^* := \mathrm{proj}_{b_+}^* a_-$, $p_+ := \mathrm{proj}_{b_+} a_+$, $c_+ := \mathrm{proj}_{b_+} c_0 \supseteq p_+$ and $\Delta_{b_+} := \{z_+ \in \Delta_+ \mid b_+ \subseteq z_+\}$. Then it follows*

$$\Gamma_{b_+} c_+ = \{x_+ \in \mathcal{C}(b_+) \mid x_+ \text{ is opposite in } \Delta_{b_+} \text{ to a chamber containing } p_+^*\} .$$

Proof: Set

$\mathcal{C}(b_+; p_+^*) := \{x_+ \in \mathcal{C}(b_+) \mid x_+ \text{ is opposite in } \Delta_{b_+} \text{ to a chamber containing } p_+^*\}$. By Proposition 4, p_+ and p_+^* are opposite in Δ_{b_+} , implying $c_+ \in \mathcal{C}(b_+; p_+^*)$. Since $\Gamma_{b_+} = G_{a_-} \cap G_{b_+}$ stabilizes Δ_{b_+} and p_+^* , it is clear that it stabilizes $\mathcal{C}(b_+; p_+^*)$ as well.

Now assume that an arbitrary $x_+ \in \mathcal{C}(b_+; p_+^*)$ is given. Since D_+ is an sfd for the action of Γ on Δ_+ and $b_+ \in D_+$, there exists a $\gamma \in \Gamma_{b_+}$ such that $y_+ := \gamma x_+ \in D_+$. Hence $y_+, c_+ \in \mathcal{C}(b_+; p_+^*) \cap \Sigma_+$, and both chambers must contain p_+ , the unique element of Σ_{b_+} opposite to p_+^* . By Lemma 4 ii), we can choose an $n \in \mathrm{Stab}_G(\Sigma)$ such that $ny_+ = c_+$. Then $np_+ = p_+ = \mathrm{proj}_{b_+} a_+$ which implies $na_+ = a_+$ by [T1], Proposition 12.5. Hence also $na_- = a_-$ and $n \in G_{a_-} \cap G_{b_+} = \Gamma_{b_+}$, showing $x_+ \in \Gamma_{b_+} c_+$. $\qquad\square$

Motivated by Lemma 12, we introduce the following

Notation: If Θ is a spherical building, $b \in \Theta$ and "op" denotes the opposition relation, we set

$$\Theta^0(b) := \{f \in \Theta \mid \text{there exist chambers } c \supseteq f \text{ and } d \supseteq b \text{ such that } c \text{ op } d\}$$

$$= \bigcup_{a \text{ op } b} \mathrm{st}_\Theta(a)$$

The complexes $\Theta^0(b)$ will be studied in detail in Chapter II. They are important for us in view of the following consequence of Lemma 11 iv) and Lemma 12:

Corollary 6: *If a_- and $b_+ \in D_+ \cap R_{j+1}$ are of spherical cotype, then*

$$T'(b_+) = \Theta^0 \left((\mathrm{proj}_{b_+}^* a_-) \setminus b_+ \right)$$

where $\Theta = \ell k_{\Delta_+}(b_+) := \{f_+ \in \Delta_+ \mid f_+ \cup b_+ \in \Delta_+ \text{ and } f_+ \cap b_+ = \emptyset\} \cong \Delta_{b_+}$. $\qquad\square$

Remark 6:

i) If Σ_+ **is locally finite**, the analysis of the homotopy properties of the filtration $(\Delta_j)_{j \geq 0}$ is reduced by Lemma 11 and Corollary 6 to the study of subcomplexes

43

of type $\Theta^0(b)$ in certain spherical buildings. In order to apply Theorem A below (see §6), we shall have to decide whether the following generalization of the Solomon–Tits theorem is true for a given spherical building Θ of rank r :

(S_Θ) $\Theta^0(b)$ is $(r-1)$–**spherical** for any $b \in \Theta$, i.e. $|\Theta^0(b)|$ is homotopy equivalent to a bouquet of $(r-1)$-spheres.

Unfortunately, (S_Θ) is definitely false for some "small" buildings and "most" of the rank 2 buildings which are not Moufang (see Ch. II, §2). It is the main goal of the subsequent Chapter II to verify (S_Θ) for every "classical" spherical building with the property that each of its panels is contained in "sufficiently many" (depending on r) chambers.

ii) If $b_+ \in D_+ \cap R_{j+1}$ is not spherical, $T'(b_+)$ cannot be described any longer without referring to the codistance δ^* . The easily proved analogue of Lemma 12 in this case is given by the equation

$$\Gamma_{b_+}(\mathrm{proj}_{b_+} c_0) = \{x_+ \in \mathcal{C}(b_+) \,|\, \exists\, c_- \in \mathcal{C}(a_-) \text{ such that } \delta^*(c_-, x_+) = w\} \,,$$

where w is minimal in $\delta^*(\mathcal{C}(a_-) \times \mathcal{C}(b_+))$.

Even more disturbing than the occurrence of δ^* is the fact that $T'(b_+)$ is hardly $(r-1)$-spherical for $r := \mathrm{rk}\; \ell k_{\Delta_+}(b_+)$. This may roughly be seen as follows: Take a subcomplex $B \subseteq \ell k_{\Delta_+}(b_+)$ not contained in $T'(b_+)$ such that $\partial B \subseteq T'(b_+)$ and the pair $(|B|, |\partial B|)$ is homotopy equivalent to the $(r-1)$-ball B^{r-1} and its boundary S^{r-2} . The existence of such a B has to be verified which, by the way, is trivial if $\ell k_{\Delta_+}(b_+)$ is a thick tree and a_- a chamber. Now by the theorem of Mayer–Vietoris, $0 \neq \widetilde{H}_{r-2}(B \cap T'(b_+))$ injects into $\widetilde{H}_{r-2}(T'(b_+))$ since $\widetilde{H}_{r-1}(\kappa) = 0$ for any subcomplex κ of $\ell k_{\Delta_+}(b_+)$ because $|\ell k_{\Delta_+}(b_+)|$ is contractible and $(r-1)$-dimensional.

In order to determine the exact "finiteness length" of the group G_{a_-} by means of Brown's criterion (see §6), we also need to establish the existence of infinitely many $j \in \mathbb{N}_0$ such that $|T(b_+)|$ is **not** contractible for some $b_+ \in R_{j+1}$. This could be done by the methods developed in Chapter II but it is more convenient to use the following

Lemma 13:   *If Σ_+ is infinite and locally finite, then there exist infinitely many chambers $c_+ \in D_+$ such that $R^\Gamma(c_+)$ is a panel. If additionally Δ_+ is thick and*

$n := \operatorname{rk} \Delta_+ - 1$, there are infinitely many $b_+ \in \left(\bigcup\limits_{j \in \mathbb{N}_0} R_{j+1} \right) \cap D_+$ such that

$\widetilde{H}_{n-1}(T(b_+)) \neq 0$.

Proof: Set $A := \{\alpha \mid \alpha$ is a root of Σ_+, $c_0 \in \alpha$ and $a_+ \in \partial\alpha\}$, and recall that $D_+ = \bigcap\limits_{\alpha \in A} \alpha$ (cf. Lemma 6 and Remark 5). Firstly, we show that there exist infinitely many vertices in the interior of D_+ , i.e. not lying on any of the walls $\partial\alpha \, (\alpha \in A)$. This is done most conveniently by using geometric arguments. It is well known that an infinite, locally finite Coxeter complex is irreducible and either of affine or of compact hyperbolic type (cf. for example [Bou2], Ch. V, §4, Exercise 14). Hence Σ_+ can be obtained from an appropriate tesselation of a Euclidean or hyperbolic space X by Euclidean or hyperbolic simplices. We identify $|\Sigma_+|$ with X and choose a ray r with origin $x_0 \in |a_+|$ such that $r \setminus \{x_0\}$ is contained in the interior of $|D_+|$. The (Euclidean or hyperbolic) distance from the points of r to any of the walls $|\partial\alpha| \, (\alpha \in A)$ is unbounded. On the other hand, every ball with radius $d := $ diameter of $|c_0|$ contains at least one vertex.

Next, we associate to every vertex v_+ in the interior of D_+ a chamber $c_+ \in D_+$ satisfying $\operatorname{rk}(R^\Gamma(c_+)) = \operatorname{rk} c_+ - 1$. One simply takes $c_+ = \operatorname{op}_{\Sigma_{v_+}}(\operatorname{proj}_{v_+} c_0)$ which is the unique chamber in Σ_{v_+} having maximal gallery-distance from c_0 . Since $c_+ \in \Sigma_{v_+} \subseteq D_+$, $R^\Gamma(c_+)$ is the minimal face b_+ of c_+ with the property $d(b_+, c_0) = d(c_+, c_0)$, and this is by construction of c_+ the panel $b_+ = c_+ \setminus v_+$ ($b_+ = c_+$ would imply the finiteness of Σ_+). This proves the first claim of the lemma. By Corollary 6, $T'(b_+)$ is the disjoint union of at least two points if Δ_+ is thick. Hence $|T(b_+)| = |\partial b_+| * T'(b_+)$ (cf. Lemma 11 iii)) contains an $(n-1)$-sphere, and the second claim follows. $\square$

§ 6 Finiteness properties of G_{a_-}

We want to determine the "finiteness length" of G_{a_-} , i.e. the largest number $m \in \mathbb{N}_0 \cup \{\infty\}$ such that G_{a_-} is of type F_m . Under suitable assumptions this can be done by applying a criterion of K.S. Brown together with the results of §5. We recall some notions which will be used in the following:

- A d-dimensional CW-complex is said to be **d-spherical** if it is $(d-1)$-connected, consequently if and only if it is contractible or homotopy equivalent to a bouquet of d-spheres.

- A d-dimensional simplicial complex is called **d-spherical** if its geometric realization is d-spherical.

- If a group Γ acts on a CW-complex by homeomorphisms which permute the cells, X is called a **Γ- CW-complex**.

As already in [Ab1], [A2] and [Ab3], a specialization of Brown's criterion will be used which is adapted to our purposes:

Lemma 14 (cf. [Br2], Corollary 3.3): *Let X be a Γ-CW-complex. Suppose that there exists an integer $n \geq 1$ such that the following conditions are satisfied:*

(a) *X is n-connected.*

(b) *If σ is a cell of dimension $d \leq n$, the stabilizer Γ_σ is of type F_{n-d} .*

(c) *$X = \bigcup\limits_{j \in \mathbb{N}_0} X_j$ with Γ-invariant subcomplexes X_j of X which are finite modulo Γ for all j .*

(d) *$X_{j+1} = X_j \cup \bigcup\limits_{i \in I_j} S_{i,j}$ with contractible subcomplexes $S_{i,j} \subseteq X_{j+1}$ satisfying*

   (d$_1$) *$S_{h,j} \cap S_{i,j} \subseteq X_j$ for all j and all $h \neq i \in I_j$*

   (d$_2$) *$S_{i,j} \cap X_j$ is $(\mathbf{n-1})$-**spherical** for all j and all $i \in I_j$*

   (d$_3$) *There exist infinitely many j such that $\widetilde{H}_{n-1}(S_{i,j} \cap X_j) \neq 0$ for at least one $i \in I_j$*

Then Γ is of type F_{n-1} and not of type FP_n . □

In view of Corollary 11 below, we also recall a well known sufficient F_m-condition which is treated in [Br2] as well (cf. loc. cit., Proposition 1.1 and Proposition 3.1).

Criterion: Let X be an $(m-1)$-connected Γ-CW-complex which is finite modulo Γ . Assume that Γ_σ is of type F_{m-d} for any cell σ of dimension $d \leq m$. Then Γ is of type F_m .

We are now going to apply Lemma 14 to the following situation:

46

$\Delta = (\Delta_+, \Delta_-, \delta^*)$ is a twin building, G a group acting strongly transitively on Δ and $\Gamma = G_{a_-}$ the stabilizer of a fixed simplex $\emptyset \neq a_- \in \Delta_-$. Then $X = |\Delta_+|$ is a Γ-CW-complex. We consider the Γ-invariant filtration of Δ_+ introduced in §5 and set $X_j := |\Delta_j|$. According to Lemma 11 i), $X_{j+1} = X_j \cup \bigcup_{b \in R_{j+1}} |S(b)|$. Hence the index set I_j is equal to R_{j+1} here, and $S_{b,j} = |S(b)|$ for $b \in I_j$.

In order to fulfill the requirements (a) – (d) of Lemma 14, we make the following assumptions:

(LF) The apartments of Δ_+ and Δ_- are infinite and locally finite.

(F) The stabilizers $\Gamma_{b_+} = G_{a_-} \cap G_{b_+}$ are finite for all $\emptyset \neq b_+ \in \Delta_+$.

(S) If Θ is the full link of a non-void simplex in Δ_+ and $d = \dim \Theta$, then $\Theta^0(x)$ is d-spherical for any $x \in \Theta$.

Discussion of the assumptions:

(LF) The set-up described above is not very fruitful if Δ_+ and Δ_- are spherical buildings. This is due to the fact that in this case $\Gamma = \Gamma_{b_+}$ for the non-void simplex $b_+ = \mathrm{proj}_{\emptyset}^* a_- \in \Delta_+$.

On the other hand, Remark 6 of §5 shows that $|S(b_+) \cap X_j| = |T(b_+)| = |\partial b_+| * |T'(b_+)|$ is hardly spherical if b_+ is not of spherical cotype. We are therefore requiring that all non-void simplices are of spherical cotype, i.e. that the apartments are locally finite. We recall once more that (LF) is satisfied precisely by those buildings which are either of irreducible affine or of compact hyperbolic type.

(F) This condition may seem to be too restrictive at first sight. But we shall see soon that it is quite natural in our context in case (LF) is satisfied. At first we state some equivalent versions of (F):

Lemma 15: *Let Δ and G be as above, assume (LF) and let $\Sigma = (\Sigma_+, \Sigma_-)$ be a twin apartment of Δ . Then the following statements are equivalent:*

i) (F)

ii) Δ_+ (and hence also Δ_-) is locally finite, and
$\mathrm{Fix}_G(\Sigma) := \{g \in G \,|\, gx = x \text{ for any } x \in \Sigma_+ \cup \Sigma_-\}$ is finite.

iii) Every panel of Δ_+ (and hence also of Δ_-) is contained in only finitely many chambers, and $\mathrm{Fix}_G(\Sigma)$ is finite.

Proof: Without loss of generality, we assume that $a_- \in \Sigma_-$.

i) $\Rightarrow$ ii): Since G acts strongly transitively on Δ ,
$st_{\Delta_+}(b_+) = \Gamma_{b_+} st_{\Sigma_+}(b_+)$ for any $b_+ \in \Sigma_+$ (cf. Lemma 2 ii), Lemma 3 and Lemma 4
iii) of §2). In view of (F) and (LF), $st_{\Delta_+}(b_+)$ is finite for any $\emptyset \neq b_+ \in \Sigma_+$. Therefore
Δ_+ is locally finite. $\mathrm{Fix}_G(\Sigma)$ is finite because it is a subgroup of Γ_{b_+} .

ii) $\Rightarrow$ iii): trivial.

iii) $\Rightarrow$ i): Choose chambers $c_+ \in \Sigma_+$, $c_- \in \Sigma_-$ such that $c_+ \mathop{op} c_-$ and $a_- \subseteq c_-$.
Let $\emptyset \neq b_+ \in \Sigma_+$ be given, and consider the action of Γ_{b_+} on $C_+ \times C_-$. We shall prove
the following two statements which obviously imply the finiteness of Γ_{b_+} :

1. The stabilizer $\mathrm{Stab}_{\Gamma_{b_+}}(c_+, c_-)$ is finite.

2. The orbit $\Gamma_{b_+}(c_+, c_-)$ is finite.

By Lemma 2 ii), an element of G stabilizing c_+ and c_- has to stabilize Σ . Hence
$\mathrm{Stab}_{\Gamma_{b_+}}(c_+, c_-) = \mathrm{Fix}_G(\Sigma)$ which is finite by assumption.

The first part of statement iii) implies the following: Given a chamber x in Δ_+ or
in Δ_- and an integer $\ell \geq 0$, there exist only finitely many chambers having gallery
distance $\leq \ell$ from x . Taking into account (LF) as well, we see that

$$\{y_+ \in C_+ \,|\, d_+(y_+, b_+) \leq d_+(c_+, b_+)\} =: M_+$$
$$\text{and} \quad \{y_- \in C_- \,|\, a_- \subseteq y_-\} =: M_-$$

are finite sets of chambers. The proof is now finished by observing
$\Gamma_{b_+}(c_+, c_-) \subseteq M_+ \times M_-$. $\qquad \square$

Corollary 7: Let $(G, (U_\alpha)_{\alpha \in \Phi}, H)$ be an RGD-system with Coxeter system $(W, S = \{s_i \mid i \in I\})$ such that W_J is finite for any $I \neq J \subset I$.

Let furthermore Δ be the twin building associated to the twin BN-pair (G, HU_+, HU_-, N, S) (cf. Proposition 1 and Example 6). Then the following holds:

i) (F) is satisfied if and only if H and all root groups U_α are finite.

ii) If $(G, (U_\alpha)_{\alpha \in \Phi}, H)$ is an RGD-system as described in one of the Examples 2 – 5 of §1, (F) is satisfied if and only if the ground field k is finite.

Proof:

i) Denote by $\Sigma = (\Sigma_+, \Sigma_-)$ the standard twin apartment of Δ containing the opposite chambers $B_+ = HU_+$ and $B_- = HU_-$.

Then $\mathrm{Fix}_G(\Sigma) = B_+ \cap B_- = H$, the latter by (RGD3). Furthermore, it is easy to see that the map

$$U_{\alpha_i} \longrightarrow \{c_+ \in C_+ \mid c_+ \text{ is } i\text{-adjacent to } B_+\}, \ u \longmapsto u\, s_i\, B_+ \ (i \in I)$$

is bijective (cf. for example [T8], Section 5, Lemma 4). Therefore, statement i) immediately follows from Lemma 15.

ii) All root groups U_α are isomorphic to the additive group of k in the Examples 2, 3 and 5. They are bijective to finite-dimensional k-vector spaces in Example 4. Hence in alle these examples, the root groups are finite if and only if k is finite. It is also easily checked that H is always finite in case k is finite. Hence statement ii) is a specialization of statement i). □

I mention in passing that in all the Examples 2 – 5 of §1, the groups G and the stabilizers G_{a_-} are even not finitely generated if k is infinite. This follows from the fact that an infinite field is never a finitely generated $\mathbb{Z}$-algebra. Hence we need not look for finiteness properties of Γ in these cases if (F) is not satisfied.

(S) This is the condition which is hardest to verify. As already announced in Remark 6, the question whether the subcomplexes of type $\Theta^0(x)$ are d-spherical for a given d-dimensional building Θ of spherical type will be treated in detail in Chapter II. Therefore, I will restrict myself to some short comments here.

1. There are some evidences indicating that the following statement is true:

   Conjecture: If Θ is a spherical Moufang building of dimension d and if every panel of Θ is contained in "sufficiently many" chambers, then $\Theta^0(x)$ is d-spherical for any $x \in \Theta$.

   In Chapter II this conjecture will be proved for all buildings of type A_{d+1} (this case is already treated in [AA]), C_{d+1} (the "exceptional" C_3 buildings described in [T1], §9, excluded) and D_{d+1} , "sufficiently many" being respectively substituted by $2^d + 1$, $2^{2d+1} + 1$ and $2^{2d+1} + 1$. These numbers result from the corresponding induction proofs. The subcomplexes $\Theta^0(x)$ are probably still spherical for significantly smaller buildings.

   The conjecture is also true for all Moufang rank 2 buildings. I am convinced that it will turn out to be correct for buildings of the exceptional types F_4, E_6, E_7 and E_8 as well. But the methods used in Chapter II will lead to technically highly complicated proofs in these cases; the argumentation is already rather involved for the D_{d+1} buildings.

2. It is easily checked (cf. Chapter II, §1, Lemma 16) that
   $\Theta^0(x) = \Theta_1^0(x_1) * \Theta_2^0(x_2)$ if $\Theta = \Theta_1 * \Theta_2$ is the join of two spherical buildings and $x = x_1 * x_2$. Hence it suffices to consider buildings with connected Coxeter diagrams in the conjecture stated above.

3. If Θ is a link in one of the buildings corresponding to the Examples 2 – 5, every panel of Θ is contained in "sufficiently many" chambers if and only if the ground field k is "big enough" (see the proof of Corollary 7).

4. If (LF) and (F) are satisfied, the (proper) links in Δ_+ are finite by Lemma 15. Hence we could restrict our attention to finite buildings in the conjecture stated above. But this would not simplify the arguments in Chapter II very much. Besides, it may be helpful in other applications if the homotopy types of the spaces $|\Theta^0(x)|$ are known for infinite spherical buildings as well.

5. At first sight, assumption (S) seems to be a purely technical condition which is needed in the proof of Theorem A. But the statement of this theorem becomes definitely wrong if one cancels assumption (S) without any substitution. This is shown by a recently discovered counter-example concerning twin buildings of compact hyperbolic type (cf. [Ab6]). It is an interesting open question whether

such counter-examples can also arise in affine situations. If so, then for small q the groups $\mathcal{G}(\mathbb{F}_q[t])$ mentioned in Corollary 9 below are possibly not always of type F_{n-1} .

We are now in a position to prove the following

Theorem A: *Assume that G is a group acting strongly transitively on a twin building $\Delta = (\Delta_+, \Delta_-, \delta^*)$, Δ_+ and Δ_- being thick n-dimensional buildings. Let a simplex $\emptyset \neq a_- \in \Delta_-$ be given, and suppose that (LF), (F) and (S) are satisfied. Then $G_{a_-} = \operatorname{Stab}_G(a_-)$ is of type F_{n-1} but not of type FP_n .*

Proof: We just have to put the pieces together. As already mentioned, the conditions (a) - (d) of Lemma 14 are to be verified for $\Gamma = G_{a_-}$, $X = |\Delta_+|$, $X_j = |\Delta_j|$ and $S_{b_+,j} = |S(b_+)|$ $(j \in \mathbb{N}_0, \ b_+ \in R_{j+1} = I_j)$.

(a) Since Δ_+ is not spherical, $|\Delta_+|$ is contractible by a straightforward generalization of the original Solomon–Tits' theorem (cf. for example [Br3], Ch. IV, §6).

(b) is of course an immediate consequence of (F).

(c) is clear by the definition of $\Delta_j = \Gamma F_j$ given in §5.

(d) $X_{j+1} = X_j \cup \bigcup_{b_+ \in R_{j+1}} |S(b_+)|$ follows from Lemma 11 i), and $|S(b_+)|$ is contractible since $S(b_+)$ is the star of b_+ in Δ_{j+1} .

(d_1) Lemma 11 ii) implies $|S(b_+)| \cap |S(b'_+)| \subseteq X_j$ for $b_+ \neq b'_+ \in R_{j+1}$.

(d_2) By definition and Lemma 11 iii),
$|S(b_+)| \cap X_j = |T(b_+)| = |\partial b_+| * |T'(b_+)|$. Since $|\partial b_+|$ is an $(s-1)$-sphere for $s := \dim b_+ = \operatorname{rk} b_+ - 1$, it remains to show that $T'(b_+)$ is $(n-s-1)$-spherical. Now Lemma 11 iv), (LF) and Corollary 6 of §5 imply $T'(b_+) = \Theta^0(x)$, where $\Theta = \ell k_{\Delta_+}(b_+)$ and $x \in \Theta$. Thus $T'(b_+)$ is $(n-s-1)$-spherical by assumption (S).

(d_3) follows from (LF) and Lemma 13. $\qquad\square$

We shall now discuss some applications of Theorem A. The first one deals with twin trees, i.e. with twin buildings of type $\tilde{A}_1$ (for a less technical definition of twin trees, the reader is referred to [RT]). Here we have $n = 1$, and assumption (S) is always satisfied for trivial reasons, 0-spherical simply meaning 0-dimensional and non-void.

Corollary 8: *Let G be a group acting strongly transitively on a thick twin tree $T = (T_+, T_-, \delta^*)$, let a_- be a vertex or an edge of T_-, and suppose that (F) is satisfied. Then G_{a_-} is not finitely generated.* □

One may suspect at once that condition (F) is superfluous in this context. This is in fact true and can be shown by the methods developed in this chapter. Beyond that, it is possible to generalize the proof of Nagao's theorem presented in [Se2], Ch. II, §1.6, to arbitrary stabilizers G_{a_-} as in Corollary 8 (recall that $SL_2(k[t])$ and also $GL_2(k[t])$ are of this form). Proposition 6 below results from combining Theorem 10 of [Se2], Ch. I, §4.5 with Proposition 3 of §3 and the Corollaries 4 and 5 of the appendix of §4. Details will be published together with the results of this appendix. However, the interested reader will have no difficulties in reproducing the proof by himself.

Proposition 6: *Let G be a group acting strongly transitively on a twin tree $T = (T_+, T_-, \delta^*)$, a_- a vertex or an edge of T_- and $\Gamma := G_{a_-}$. Choose a twin apartment $\Sigma = (\Sigma_+, \Sigma_-)$ of T such that $a_- \in \Sigma_-$, and denote by a_+ the simplex of Σ_+ opposite to a_-. Number the vertices of Σ_+ in the form $\ldots x_{-2}, x_{-1}, x_0, x_1, x_2, \ldots$ such that $\{x_j, x_{j+1}\}$ is an edge and $x_j \neq x_{j+2}$ for all $j \in \mathbb{Z}$. Assume additionally that either $a_+ = x_0$ or $a_+ = \{x_{-1}, x_0\}$. Setting $\Gamma_j := \Gamma_{x_j}$ for $j \in \mathbb{Z}$, one obtains:*

i) *If a_- is a vertex, $\Gamma = \Gamma_0 *_{\Gamma_0 \cap \Gamma_1} (\overset{\infty}{\underset{j=1}{\cup}} \Gamma_j)$ with $\Gamma_j \subseteq \Gamma_{j+1}$ for all $j \in \mathbb{N}$.*

ii) *If a_- is an edge, $\Gamma = (\overset{-\infty}{\underset{j=-1}{\cup}} \Gamma_j) *_{\Gamma_{-1} \cap \Gamma_0} (\overset{\infty}{\underset{j=0}{\cup}} \Gamma_j)$ with $\Gamma_j \subseteq \Gamma_{j+1}$ for all $j \in \mathbb{N}_0$ and $\Gamma_{-j} \subseteq \Gamma_{-j-1}$ for all $j \in \mathbb{N}$.*

iii) *If T_+ is thick, $\Gamma_j \neq \Gamma_{j+1}$ for all $j \in \mathbb{Z}$, and Γ is not finitely generated.*

I mention in passing that the unions occurring in i) and ii) are the stabilizers in Γ of the corresponding ends of Σ_+. Note also that these two statements can be applied to G as well since $G = G_{x_-} *_{G_{a_-}} G_{y_-}$ for any edge $a_- = \{x_-, y_-\}$ of T_-.

We are now turning to further applications of Theorem A. They are concerned with the examples treated in §1. If Ψ' is a reduced root system, $\mathcal{G}'$ a Chevalley group of type Ψ' and k a field, the spherical building $\Delta(\mathcal{G}', k)$ of $\mathcal{G}'$ over k (cf. [T1], §5) is (up to isomorphism) independent of the choice of $\mathcal{G}'$. It will be denoted by $\Delta(\Psi', k)$ in the following. Let Ψ and $\mathcal{G}$ be given as in Example 3, and let Δ_+ be the Bruhat–Tits building associated to $(\mathcal{G}(k[t,t^{-1}])^+, B_+, N, S)$. Then all links in Δ_+ are of the form $\Delta(\Psi', k)$, where Ψ' is a root system with Dynkin diagram diag(Ψ') which is a proper subdiagram of the extended Dynkin diagram $(\Psi)^\sim$. This fact is well known; cf. [Ab3], Corollary 2, for a proof, if necessary.

In view of Corollary 7, we shall assume that $k = \mathbb{F}_q$ is finite in the following. Since $\mathcal{G}(\mathbb{F}_q[t])^+$, being maximal parabolic with respect to $(\mathcal{G}(\mathbb{F}_q[t, t^{-1}]^+, B_-, N, S)$, is the stabilizer of a vertex in Δ_-, Theorem A implies

Corollary 9: *Let Ψ be a reduced and irreducible root system of rank n and $\mathcal{G}$ a Chevalley group of type Ψ. Assume that $\Theta^0(x)$ is $(\ell - 1)$-spherical for all root systems Ψ' of rank $\ell \leq n$ with diag(Ψ') $\subset$ diag(Ψ)$^\sim$ and all $x \in \Theta := \Delta(\Psi', \mathbb{F}_q)$. Then $\mathcal{G}(\mathbb{F}_q[t])^+$ — and hence also $\mathcal{G}(\mathbb{F}_q[t])$ since $[\mathcal{G}(\mathbb{F}_q[t]) : \mathcal{G}(\mathbb{F}_q[t])^+] < \infty$ — is of type F_{n-1} but not of type FP_n.* □

This is the main result of [Ab3]. A similar statement is true if $\mathcal{G}$ is an almost simple, isotropic $\mathbb{F}_q$-group. We shall discuss this in Chapter III for those cases where assumption (S) can be verfied by using the results of Chapter II.

Generalizing Corollary 9 in a different direction, we shall consider **Kac–Moody groups over finite fields** next. Let $D, \mathcal{G}_D, \mathcal{T}, \Phi, (\mathcal{U}_\alpha)_{\alpha \in \Phi}$ be as in Example 5 and set $G := \mathcal{G}_D(\mathbb{F}_q)$, $H := \mathcal{T}(\mathbb{F}_q)$, $U_\alpha := \mathcal{U}_\alpha(\mathbb{F}_q)$ ($\alpha \in \Phi$).
Let $(G, B_+ = HU_+, B_- = HU_-, N, S)$ be the twin BN-pair corresponding to the RGD-system $(G, (U_\alpha)_{\alpha \in \Phi}, H)$ and $(\Delta_+, \Delta_-, \delta^*)$ the twin building associated to it.

If (LF) is satisfied, every link in Δ_+ is a finite Moufang building with the property that each of its panels is contained in exactly $q+1$ chambers (cf. the proof of Corollary 7). In view of the classification of finite Moufang buildings (cf. [T1], §11, [FS] and [Ro], Appendix 6), each of these links has to be of the form $\Delta(\Psi', \mathbb{F}_q)$. The links in Δ_+ are therefore completely determined by their Coxeter diagrams with the only exception that $\Delta(B_\ell, \mathbb{F}_q)$ cannot be distinguished from $\Delta(C_\ell, \mathbb{F}_q)$ in this way. In order to achieve this as well, one has to analyse the full Kac–Moody data D . However, since $\Delta(B_\ell, \mathbb{F}_q)$ and $\Delta(C_\ell, \mathbb{F}_q)$ have similar properties from our point of view, we shall not discuss these technicalities here.

Summing up, we obtain the following consequence of Theorem A:

Corollary 10: *Let $\mathcal{G}_D$ be a Kac–Moody group functor, (W, S) the Coxeter system associated to D, $n := \#S - 1$ and Γ a proper subgroup of $G = \mathcal{G}_D(\mathbb{F}_q)$ which is parabolic with respect to (G, B_-, N, S) . Suppose that the following two conditions are satisfied:*

1. *(W, S) is of irreducible affine or of compact hyperbolic type.*

2. *If the Coxeter diagram underlying $\mathrm{diag}(\Psi')$ is contained in the Coxeter diagram of (W, S), $\Theta^0(x)$ is $(\ell - 1)$-spherical for $\Theta := \Delta(\Psi', \mathbb{F}_q)$, $\ell := \mathrm{rk}\,\Psi'$ and any $x \in \Theta$.*

Then Γ is of type F_{n-1} but not of type FP_n . $\quad\square$

Combining the criterion below Lemma 14 with Theorem A, we also obtain a result concerning the full group G . Note that $\bar{c}_- = \{a_- \in \Delta_- \mid a_- \subseteq c_-\}$ is an sfd for the action of G on Δ_- for every chamber $c_- \in \Delta_-$.

Corollary 11: *Assume that G and Δ are given as in Theorem A . Suppose further that (LF), (F) and (S) are satisfied. Then G is of type F_{n-1} .* $\quad\square$

Remark 7: I am convinced that Corollary 11 only describes half of the truth. The complete result I expect is that G is of type F_{2n-1} but not of type FP_{2n} . This can presumably be deduced — appropriate conditions again presupposed — from properties of the action of G on the polysimplicial complex $\Delta_+ \times \Delta_-$. However, two new problems are arising here. Firstly, this action does not admit a (polysimplicial) fundamental domain in the strict sence used in §5 though $\Sigma_+ \times \bar{c}_-$ comes close to being one according to Proposition 2 of §3. Secondly, the relative links occurring with G-invariant filtrations of $\Delta_+ \times \Delta_-$ are significantly different from the complexes $\Theta^0(x)$ introduced above. Nevertheless, I hope that a program roughly analogous to that described for the stabilizers G_{a_-} can be carried out for the group G as well.

Theorem A and all its corollaries (with the exception of Corollary 8) depend on the assumption (S). In the discussion above, I already announced some results concerning this condition. The corresponding proofs should not be postponed any longer. So let us enter the realm of spherical buildings now.

II   Homotopy properties of $|\Delta^0(a)|$

§ 1   Basic properties of $\Delta^0(a)$

This chapter is devoted to the analysis of topological properties of $|\Delta^0(a)|$, where Δ is a spherical building, $a \in \Delta$ and $\Delta^0(a)$ the subcomplex of Δ introduced in Chapter I, §5. (Since we shall not consider twin buildings in this context, the letter Δ will always be reserved for **spherical buildings** from now on.) We shall show that $|\Delta^0(a)|$ has the homotopy type of a bouquet of $(\dim \Delta)$-dimensional spheres provided that Δ is Moufang, "big enough" and no irreducible factor of Δ is associated to a k-form of one of the exceptional groups. It is this result which makes it possible to verify the crucial condition (S) of the previous chapter in certain cases, thereby yielding applications of Theorem A which do not depend any longer on any assumptions.

The proof of the above statement concerning $|\Delta^0(a)|$ goes by induction on rk Δ and uses the description of "classical" spherical buildings as flag complexes of certain geometries. A similar approach should work in the case of "exceptional" buildings as well. However, since the corresponding explicit descriptions are technically complicated, necessitating also the introduction of a bunch of further notations (and since the D_n-case is already sufficiently involved), the exceptional buildings will not be treated in this book.

Though we shall have to deal with buildings of type A_n, C_n and D_n separately in §§4 – 7 below, some basic facts concerning $\Delta^0(a)$ can be derived from the general theory of spherical buildings without specifying the type of Δ , cf. [Ab4]. For the convenience of the reader I shall include short, modulo [T1] self-contained proofs of those statements of [Ab4], §1.1, which are used in the following. We denote by op the opposition relation in Δ and, for any apartment Σ of Δ , by $\mathrm{op}_\Sigma : \Sigma \longrightarrow \Sigma$ the opposition involution of Σ . Recall that

$$\Delta^0(a) \ := \ \{f \in \Delta \,|\, \text{there exist chambers } c \supseteq f \text{ and } d \supseteq a \text{ such that } c \text{ op } d\}$$

$$= \ \bigcup_{b \,\mathrm{op}\, a} \mathrm{st}_\Delta(b)$$

Lemma 16:   *Given $a \in \Delta$ and $\Delta^0(a)$ as above, one obtains:*

i) If Σ is any apartment containing a and $a^0 := \mathrm{op}_\Sigma(a)$, then
$$\Delta^0(a) \cap \Sigma = \Sigma^0(a) = \mathrm{st}_\Sigma(a^0) .$$

ii) $\Delta^0(a)$ is a full subcomplex of Δ , i.e. $e \cup f \in \Delta^0(a)$ for any two $e, f \in \Delta^0(a)$ such that $e \cup f \in \Delta$.

iii) If $a, b, a \cup b \in \Delta$, then $\Delta^0(a \cup b) = \Delta^0(a) \cap \Delta^0(b)$.

iv) If $\Delta = \Delta_1 * \Delta_2$ is the join of two spherical buildings Δ_1, Δ_2 and $a = a_1 * a_2$, then $\Delta^0(a) = \Delta^0(a_1) * \Delta^0(a_2)$.

Proof:

i) It is clear that $\Sigma^0(a) = \mathrm{st}_\Sigma(a^0) \subseteq \Delta^0(a) \cap \Sigma$. So let $f \in \Delta^0(a) \cap \Sigma$ and opposite chambers $c, d \in \Delta$ satisfying $c \supseteq f$, $d \supseteq a$ be given. Denote by Σ' the (unique) apartment containing c and d and choose an isomorphism $\alpha : \Sigma' \xrightarrow{\sim} \Sigma$ fixing f and a . Then $\alpha(c)$ and $\alpha(d)$ are opposite chambers of Σ containing f and a , respectively. Hence $f \in \Sigma^0(a)$.

ii) Choose an apartment Σ of Δ containing a and $e \cup f$. By i), $e \cup a^0$ and $f \cup a^0$ are elements of Σ for $a^0 = \mathrm{op}_\Sigma(a)$. Hence $e \cup f \cup a^0 \in \Sigma$ as well by [T1], Corollary 2.27. This implies $e \cup f \in \mathrm{st}_\Sigma(a^0) \subseteq \Delta^0(a)$.

iii) We have to show $\Delta^0(a) \cap \Delta^0(b) \subseteq \Delta^0(a \cup b)$. So let $f \in \Delta^0(a) \cap \Delta^0(b)$ be given. By definition, this also means $a \in \Delta^0(f)$ and $b \in \Delta^0(f)$. Therefore $a \cup b \in \Delta^0(f)$ by ii), implying $f \in \Delta^0(a \cup b)$.

iv) Given chambers $c_1, d_1 \in \Delta_1$ and $c_2, d_2 \in \Delta_2$, the gallery-distance $d_\Delta(c, d)$ between $c := c_1 * c_2$ and $d := d_1 * d_2$ in Δ is equal to the sum $d_{\Delta_1}(c_1, d_1) + d_{\Delta_2}(c_2, d_2)$. Hence c and d are opposite in Δ if and only if c_i and d_i are opposite in Δ_i for $i = 1, 2$. Combined with the definition of $\Delta^0(a)$, this implies $\Delta^0(a) = \Delta^0(a_1) * \Delta^0(a_2)$. $\square$

Lemma 17: *Assume that a group G acts strongly transitively (in the usual sense, cf. [Br2], Ch. V) on Δ . Then the following holds:*

i) *For any $a \in \Delta$, $\mathrm{Stab}_G(a)$ acts transitively on the set of chambers $\mathcal{C}(\Delta^0(a))$.*

ii) *Let c be a chamber of Δ and Σ an apartment containing c . Set*
   $B := \mathrm{Stab}_G(c)$ and $H := \mathrm{Fix}_G(\Sigma) = \bigcap_{c' \in \mathcal{C}(\Sigma)} \mathrm{Stab}_G(c')$. Assume that there
   exists a subgroup U of B satisfying $B = UH$ and $U \cap H = \{1\}$. Then U acts
   simply transitively on $\mathcal{C}(\Delta^0(c))$.

Proof:

i) Since G acts strongly transitively on Δ , $\mathrm{Stab}_G(a)$ acts transitively on all apartments containing a and hence also on all simplices opposite to a . Let b be such
   a simplex and c_1, c_2 two chambers containing b. Since a is opposite to b , there
   exists an apartment Σ containing c_1, c_2 and a (take an apartment containing c_2
   and $\mathrm{proj}_a c_1$ and note that $\mathrm{proj}_b (\mathrm{proj}_a c_1) = c_1$ by [T1], Theorem 3.28). Again
   by strong transitivity, there is an $n \in \mathrm{Stab}_G(\Sigma)$ satisfying $nc_1 = c_2$. Since the
   action of G is type-preserving, this implies $nb = b$. Finally $na = a$, because a
   is the unique simplex of Σ opposite to b .

ii) Applying i), one obtains $\mathcal{C}(\Delta^0(c)) = Bc^0 = UHc^0 = Uc^0$, where $c^0 := \mathrm{op}_\Sigma(c)$.
   Since Σ is the unique apartment containing c and c^0 ,
   $\mathrm{Stab}_G(c) \cap \mathrm{Stab}_G(c^0) = \mathrm{Fix}_G(\Sigma)$. Therefore, $uc^0 = c^0$ and $u \in U$ imply
   $u \in U \cap H$ and hence $u = 1$. □

Remark 8:

i) Sometimes a "small" subgroup of $\mathrm{Stab}_G(a)$ still acts transitively on $\mathcal{C}(\Delta^0(a))$.
   For example, if Δ is the spherical building Δ_{b_+} occurring in Lemma 12 of
   Chapter I, 5, a the simplex $\mathrm{proj}^*_{b_+} a_-$ and G the group G_{b_+} , then not only
   $G_a = \mathrm{Stab}_G(a)$ but even $\Gamma_{b_+} = G_{b_+} \cap G_{a_-}$ acts transitively on $\mathcal{C}(\Delta^0(a))$ according to that lemma. For this reason we are studying the simplicial complexes
   $\Delta^0(a)$ (and not some of their subcomplexes) here.

ii) The canonical examples satisfying all the assumptions made in Lemma 17 ii)
   arise in the following way: Let G be a group, Φ a (not necessarily reduced) root
   system and $(H, (U_\alpha)_{\alpha \in \Phi})$ a generating root datum in G in the sense of [BrT1],
   §6.1. Define U to be the subgroup of G generated by all U_α with $\alpha \in \Phi_+$,
   where Φ_+ is the set of all positive roots with respect to a fixed base Π of Φ .
   Recall that G possesses a BN-pair with $B = UH$, and consider now the action
   of G on the spherical building $\Delta = \Delta(G, B)$.

§ 2 Examples and counter-examples

Before treating the classical spherical buildings in detail, we consider some (classes of) examples in order to see whether $\Delta^0(a)$ is $(\dim \Delta)$-spherical or not in these cases. Here we shall concentrate on thick buildings Δ of low rank and on subcomplexes $\Delta^0(a)$ where a is a chamber. The results discussed in this section are technically not needed in the following. However, they are interesting in their own right. Proposition 9 for example answers an open question concerning generalized polygons which is indicated in [T7], Section 16, and explicitly formulated in [Brou], Section 5. In the present section, the proofs are just sketched, but the interested reader can easily complete them by following the main arguments outlined in the text.

The rank 2 case

Here we are dealing with the question whether the subgraph $\Delta^0(a)$ of a given **generalized m-gon** Δ (i.e. of a building of type $\bullet \overset{m}{\rule{1cm}{0.4pt}} \bullet$, cf. [Ro], Ch. 3, §2) is connected. We distinguish three cases, always assuming that Δ is thick.

a) Δ is Moufang

In this case, the problem can be transferred into a purely group theoretic one. Let Σ be an apartment of Δ . Number the vertices $v_1, \ldots, v_{2m}$ of Σ such that $\{v_1, v_2\}, \{v_2, v_3\}, \ldots, \{v_{2m}, v_1\}$ are the chambers of Σ . Denote by α_i the root of Σ containing $v_i, v_{i+1}, \ldots, v_{i+m}$ and by $U_i := U_{\alpha_i}$ the corresponding root group (cf. [Ro], Ch. 6). Set

$$G := \langle U_i \, | \, 1 \leq i \leq 2m \rangle \leq \mathrm{Aut}(\Delta) \, , \quad U := \langle U_i \, | \, 1 \leq i \leq m \rangle \, ,$$
$$H := \bigcap_{i=1}^{2m} N_G(U_i) \, , \quad c := \{v_m, v_{m+1}\} \text{ and } B := \mathrm{Stab}_G(c) \, .$$

Then the following facts are well known (cf. [T5], Section 2, or again [Ro], Ch. 6):

- $H = \mathrm{Fix}_G(\Sigma)$, and B is the semidirect product of H and U

- B is part of a BN-pair for G , and the corresponding building $\Delta(G, B)$ is isomorphic to Δ

- the product map $U_1 \times U_2 \times \ldots \times U_m \longrightarrow U$ is bijective and
  $[U_1, U_m] \leq U_2 \ldots U_{m-1}$

In particular, we can apply Lemma 17 ii) and obtain $\mathcal{C}(\Delta^0(c)) = Uc^0$ for $c^0 := \{v_{2m}, v_1\}$. It is easily deduced from the properties of the root groups U_1 and U_m that c^0 and uc^0 ($u \in U$) can be connected by a gallery in $\Delta^0(c)$ if and only if

$$u \in U' := \langle U_1, U_m \rangle = U_1[U_1, U_m]U_m \leq U$$

Hence we obtain the following

Lemma 18: *The index $[U : U']$ is equal to the number of connected components of $\Delta^0(c)$. In particular, $\Delta^0(c)$ is connected if and only if $[U_1, U_m] = U_2 \ldots U_{m-1}$.*

□

Using some parts of Tits' classification of Moufang m-gons, it is not difficult to list all cases where $U' \neq U$:

Proposition 7: *Let Δ be a thick Moufang m-gon not associated to one of the following 4 groups:*

$$C_2(\mathbb{F}_2) = Sp_4(\mathbb{F}_2), \quad G_2(\mathbb{F}_2), \quad G_2(\mathbb{F}_3), \quad {}^2F_4(\mathbb{F}_2)$$

Then $\Delta^0(c)$ is connected.

Sketch of proof: According to the theorem of Tits and Weiss, $m \in \{3, 4, 6, 8\}$.

m=3: $[U_1, U_3] = U_2$ is well known (cf. [T5], Corollary 2.10).

m=4: Setting $U_i^* := U_i \setminus \{1\}$, $[U_1^*, U_4^*] = U_2^* U_3^*$ follows from [T5], Proposition 2.9.

m=6: We use the following two facts concerning Moufang hexagons:

1. The root groups $U_1, \ldots, U_{12}$ of a Moufang hexagon constitute a root datum of type G_2 in the sense of [BrT1] (cf. [T12] for a proof).

2. Assuming 1., it is shown in [F], Chapter 3, that there exists a (commutative) field k and a Jordan division algebra J over k satisfying the following conditions:

   i) The root groups corresponding to the long roots of G_2 are coordinatized by the additive group of k.

   ii) The root groups corresponding to the short roots of G_2 are coordinatized by the additive group of J.

iii) The commutator formulae expressing $[u_i, u_j]$ for $u_i \in U_i$, $u_j \in U_j$, $i < j < i + 6$ as an element of $U_{i+1} \ldots U_{j-1}$ are given by 1. (if $[U_i, U_j] = \{1\}$) and by Theorem 3.55 in [F].

Analyzing these commutator formulae, one obtains $U' = U$ with the following two exceptions: $J = k = \mathbb{F}_2$ and $J = k = \mathbb{F}_3$. In these cases, $[U : U'] = 4$ and $[U : U'] = 3$, respectively.

m=8: According to the classification of Moufang octagons (cf. [T5]), Δ corresponds to a Ree group of type $^2F_4(k, \sigma)$, where k is a field of characteristic 2 and σ an endomorphism of k satisfying $\sigma^2(\lambda) = \lambda^2$ for all $\lambda \in k$. In particular, $U = U(k)$ is a subgroup of $^2F_4(k, \sigma)$ and $U(\mathbb{F}_2)$ is canonically embedded in U. This leads to the following distinction of cases:

1. $k = \mathbb{F}_2$ The commutator formulae stated in [T5], §1.7.1, are easy to handle here and make an explicit calculation of U' possible. Using Tits' notations, the result is the following:

$$U'(\mathbb{F}_2) = \{t_1 \cdot t_2 \cdot t_{2'} \cdot \ldots \cdot t_8 \cdot t_{8'} \,|\, t_i, t_{2j'} \in \mathbb{F}_2; \; t_2 + t_4 + t_6 = 0\}$$

2. $k \neq \mathbb{F}_2$ Applying 1. and again [T5], §1.7.1, one deduces $U'_2, U'_4, U'_6, U_3, U_5, U_7 \subseteq U' = U'(k)$. Now the formula for $[t_1, u_8]$ implies $x_2 y_4 \in U'$ for all $x, y \in k^*$. Putting the pieces together, we obtain $U' = U$ for $k \neq \mathbb{F}_2$. □

Remark 9:

i) The classification of Moufang hexagons is also due to Tits. It is stated without proof in [T2]. However, those parts of this classification which are needed above are completely proved in the quoted publications.

ii) In [T7], §16.7, Tits mentions without proof a statement which is essentially equivalent to Proposition 7. Only the Moufang octagons are not considered in that context and the counter-example $G_2(\mathbb{F}_3)$ is missing. The latter was observed by Baumgartner. Using the classification of finite BN-pairs of rank 2 (cf. [FS]), he gave a detailed proof of Proposition 7 for finite Moufang m-gons with $m \neq 8$ in [Ba], appendix.

iii) Clearly $\Delta^0(a)$ is connected for every vertex $a \subset c$ if $\Delta^0(c)$ is connected. In the four cases excluded in Proposition 7, we obtain by a reasoning similar to Lemma 18: $\Delta^0(a)$ is not connected if and only if Δ is associated to $G_2(\mathbb{F}_2)$ or to $^2F_4(\mathbb{F}_2)$ and a corresponds to a short root. In these cases, $\Delta^0(a)$ possesses exactly 2 connected components (cf. also [Brou]).

b) Δ is finite

Using finite graph theory, Brouwer was able to prove that $\Delta^0(c)$ $(c \in \mathcal{C}(\Delta))$ is connected for "almost all" finite m-gons. The counter-examples turn out to be exactly those mentioned in Proposition 7, possibly together with non-classical hexagons of order $(3,3)$ and non-classical octagons of order $(2,4)$ or $(4,2)$ if these exist. Recall that a generalized m-gon is said to be of order (s,t) if every vertex of the first type has valency $s + 1$ and every vertex of the second type has valency $t + 1$.

Proposition 8 (cf. [Brou], Theorem 1.1): *If Δ is a thick finite m-gon of order (s,t) and $(m,s,t) \notin \{(4,2,2),(6,2,2),(6,3,3),(8,2,4),(8,4,2)\}$, then $\Delta^0(c)$ is connected for any chamber $c \in \Delta$.* $\qquad\square$

c) Δ is an arbitrary (thick) m-gon

So neither the tools of group theory nor those of finite graph theory can be applied in order to decide whether $\Delta^0(c)$ $(c \in \mathcal{C}(\Delta))$ is connected or not. Nevertheless, it is still possible to describe what happens in the general case.

$m=2$: Obviously $\Delta^0(c)$ is connected.

$m=3$: It is easy to check that $\Delta^0(c)$ is connected.

$m=4$: One can show that $\Delta^0(c)$ is connected provided that Δ is not of order $(2,2)$. For example, the arguments mentioned in [Brou], Section 5, can be organized such that they constitute a proof of this fact.

$m \geq 5$: Here I expected for a long time similar results as above, i.e. connectedness of $\Delta^0(c)$ in the "generic" case and exceptions only for "small" buildings. However, since this was hard to prove, I started to look for new counter-examples instead. Having changed my expectations, it was not very difficult to deduce the following

Proposition 9: *For every integer $m \geq 5$, there exists a generalized m-gon Δ together with a chamber $c \in \Delta$ satisfying the following conditions:*

i) *Every vertex of Δ is contained in infinitely many chambers.*

ii) *$\Delta^0(c)$ is an infinite disjoint union of trees.*

Sketch of proof: The key idea consists in modifying appropriately the "free construction" described in [T3], §4.4. Take an arbitrary connected, bipartite graph of girth ($:=$ length of a shortest cycle) $2m$ and fix an edge c of Γ. Construct a bipartite graph Γ' containing Γ as follows (d_Γ denotes the usual distance in Γ): If x and y are vertices of Γ with $d_\Gamma(x,y) = m + 1$, insert a new path $p(x,y)$ of length $m - 1$ connecting them. If additionally $d_\Gamma(x,c) \geq m - 1$ and $d_\Gamma(y,c) \geq m - 1$, insert a further path $p(x,y,c)$ of length $m - 2$ connecting a central vertex $z(x,y)$ of $p(x,y)$ (for even m, there are two possibilities) with c. This is done such that the type function on $\Gamma \cup p(x,y)$ can be extended to $\Gamma \cup p(x,y) \cup p(x,y,c)$.

Example: $m = 5$ $p(x,y)$:

$$z := z(x,y)$$

Now we define a sequence of graphs by $\Gamma_{i+1} := \Gamma'_i$ ($i \in \mathbb{N}_0$) and denote by $\Gamma_i^0(c)$ the subgraph of Γ_i generated by all vertices $z \in \Gamma_i$ such that $d_{\Gamma_i}(z,c) \geq m - 1$. Then the following facts can be verified:

1. $\Delta := \overset{\infty}{\underset{i=0}{\bigcup}} \Gamma_i$ is an m-gon.

2. If every vertex of Γ_0 is contained in at least 3 edges of Γ_1, then Δ is thick.

3. $\Delta^0(c) = \overset{\infty}{\underset{i=0}{\bigcup}} \Gamma_i^0(c)$.

4. Elements of different connected components of $\Gamma^0(c)$ lie in different connected components of $\Delta^0(c)$.

5. Every cycle of $\Delta^0(c)$ is contained in $\Gamma^0(c)$.

63

It is now easy to find a graph Γ such that Δ satisfies i) and ii). Take for example $\Gamma = \bigcup_{j=1}^{\infty} C_j$, where each C_j is a $2m$-cycle and $C_j \cap C_{j'} = c$ for all $j \neq j'$. $\square$

In view of the freedom one has to construct further counter-examples, it is definitely not possible to say that $\Delta^0(c)$ is connected "in the generic case" (whatever this means precisely) for non-Moufang m-gons with $m \geq 5$.

The rank 3 case

Having already discussed the rank 2 buildings, we may assume in view of Lemma 16 iv) that Δ is of irreducible type, i.e. that it is either an A_3 or a C_3 building. According to Tits' classification of thick spherical buildings (cf. [T1]), Δ is the building associated to a root datum of type A_3, B_3, C_3 or BC_3. This allows to translate again questions concerning topological properties of $|\Delta^0(c)|$ ($c \in \mathcal{C}(\Delta)$) into group theoretic problems.

Let $(H, (U_\alpha)_{\alpha \in \Phi})$ be a root datum corresponding to Δ (cf. Remark 8 ii)), $\Pi = \{\alpha_1, \alpha_2, \alpha_3\}$ a base of Φ and Φ_+ the associated set of positive roots.

Set $U \quad := \langle U_\alpha \,|\, \alpha \in \Phi_+\rangle, \quad U_i := U_{\alpha_i} \ (1 \leq i \leq 3)$ and

$\quad U_{ij} \quad := \langle U_{p\alpha_i + q\alpha_j} \,|\, p, q \in \mathbb{N}_0 \,;\, p\alpha_i + q\alpha_j \in \Phi\rangle \ (1 \leq i \neq j \leq 3)$.

Denote by $\tilde{U}$ the amalgamated product of U_{12}, U_{13}, U_{23} with respect to their intersections $U_1 = U_{12} \cap U_{13}, \ U_2 = U_{12} \cap U_{23}, \ U_3 = U_{13} \cap U_{23}$.

We now fix the chamber $c = B = UH \in \Delta = \Delta(G, B)$. It is easy to verify $U = \langle U_{12}, U_{13}, U_{23}\rangle$, which implies that $|\Delta^0(c)|$ is connected. By Proposition 6 in [T7], $|\Delta^0(c)|$ is simply connected if and only if the canonical homomorphism $\tilde{U} \longrightarrow U$ is an isomorphism. In view of Lemma 17 ii), Theorem 1.1 in [Sw] implies more precisely:

Lemma 19: *There exists an exact sequence of the form*

$$1 \longrightarrow \pi_1(|\Delta^0(c)|) \longrightarrow \tilde{U} \longrightarrow U \longrightarrow 1$$

$\square$

We shall discuss the cases A_3 and C_3 separately now.

a) Δ is of type A_3

This case is completely treated in [T7], Section 16. The result is the following:

Proposition 10: *Let Δ be a thick building of type A_3 .*

i) *If $\Delta \ncong \Delta(A_3, \mathbb{F}_2)$, then $\tilde{U} \cong U$ and $\Delta^0(c)$ is 2-spherical.*

ii) *If $\Delta \cong \Delta(A_3, \mathbb{F}_2)$, then $\pi_1(|\Delta^0(c)|) \cong \mathbb{Z} \times \mathbb{Z}$, and $|\Delta^0(c)|$ is homeomorphic to the torus $S^1 \times S^1$.*   □

By the way, almost the same result (sphericity of $\Delta^0(c)$ in case $\Delta = \Delta(A_3, k)$ and $\#k \geq 4$) can be derived by the geometric/combinatorial approach introduced in [AA] in order to study the A_n buildings of higher rank (cf. also §4 below).

b) Δ is of type C_3

First we generalize some arguments used in [T7], Section 16, in the A_3 case. We may assume that the numbering $\{\alpha_1, \alpha_2, \alpha_3\}$ corresponds to the diagram $\overset{\textstyle 1 \quad\; 2 \quad\; 3}{\bullet\!\!-\!\!\!-\!\!\bullet\!\!-\!\!\!-\!\!\bullet}$

Set $X_i := \langle U_{p\alpha_i + q\alpha_2} \mid p \in \mathbb{N}, q \in \mathbb{N}_0 ; \, p\alpha_i + q\alpha_2 \in \Phi \rangle$ for $i = 1, 3$.

Then $U_{i2} = U_2 \ltimes X_i$ $(i = 1, 3)$, whereas $U_{13} = U_1 \times U_3$. Denote by R_i the set of all relations in X_i , and set ${}^u x := uxu^{-1}$. Rearranging the presentation of $\tilde{U}$, we get $\tilde{U} = U_2 \ltimes \langle X_1 \cup X_3 \mid R_1; R_3; [{}^{u_2}u_1, {}^{u_2}u_3] = 1, u_j \in U_j, 1 \leq j \leq 3 \rangle$. Denote the second factor on the right side of this equation by X . Applying Lemma 19, one obtains some further counter-examples now. The isometry groups occurring below are denoted as in [HO], §6.2.E. In finite group theory, different notations (such as $U_6(2)$ instead of $U_6(\mathbb{F}_4)$) are more common.

Example 8:   **The buildings associated to $\mathbf{Sp_6(\mathbb{F}_2), Sp_6(\mathbb{F}_3), U_6(\mathbb{F}_4)}$ and ${}^2\mathbf{O_8(\mathbb{F}_2)}$**

We shall show in all these cases that $\pi_1(|\Delta^0(c)|)$ is infinite or, what amounts to the same, that the group X introduced above is infinite. Note that $X_1 = U_1 \times U_{\alpha_1 + \alpha_2} = U_1 \times^u U_1$ for any $u \in U_2 \setminus \{1\}$.

i) If we are dealing with the group $Sp_6(\mathbb{F}_3)$ or with $U_6(\mathbb{F}_4)$, it is easy to verify
$$X_3 = \underset{u \in U_2}{\times} {}^u U_3 \, .$$
Hence for any $u \in U_2 \setminus \{1\}$, X projects onto the infinite group
$$\underset{v \neq 1, u}{\times} \, {}^v U_3 \times (U_1 * {}^u U_3) \times ({}^u U_1 * U_3) \, .$$

ii) Consider now either $Sp_6(\mathbb{F}_2)$ or the non-split orthogonal group ${}^2O_8(\mathbb{F}_2)$ corresponding to a quadratic $\mathbb{F}_2$-space of dimension 8 and Witt index 3. Set $X_3' := X_{2\alpha_2 + \alpha_3}$ in the first case and $X_3' := X_{\alpha_2 + 2\alpha_3}$ in the second. Both times we obtain $X_3/X_3' \cong U_3 \times {}^u U_3$ for the unique element $u \in U_2 \setminus \{1\}$. Setting $X_3' = 1$, we see that X projects onto $(U_1 * {}^u U_3) \times ({}^u U_1 * U_3)$.

Using some computer calculations, one can also show that $\tilde{U}$ (though possibly finite) is strictly bigger than U in the two cases corresponding to $O_7(\mathbb{F}_3)$ and to $Sp_6(\mathbb{F}_4)$. I do not expect any further counter-examples in the C_3 case. However, it is a tedious job to deduce all necessary relations for U from the presentation of $\tilde{U}$. Instead, one can use the method of §6 in order to prove that $\Delta^0(c)$ is 2-spherical for "most" C_3 buildings. Analysing the proof (not just the statement; cf. also Remark 14) of Proposition 13 in §6, one obtains the following result:

Proposition 11: *Assume that Δ is a thick building of type C_3.*

i) *If Δ is infinite and corresponds to a polar space with Desarguesian planes (cf. [T1], §8), then $\Delta^0(c)$ is 2-spherical.*

ii) *If Δ is associated to $Sp_6(\mathbb{F}_q), O_7(\mathbb{F}_q), {}^2O_8(\mathbb{F}_q), U_6(\mathbb{F}_q), U_7(\mathbb{F}_q)$, respectively, then $\Delta^0(c)$ is 2-spherical provided that $q > 11, 13, 11, 25, 9$, respectively.* □

I hope to complete both parts of this proposition in the future. On the one hand, one has to consider additionally the non-embeddable polar spaces discussed in [T1], §9. On the other side, the unpleasant task of checking the relations for U should be carried out by a computer in those few cases not covered by Example 8 and by Proposition 11 ii).

Higher ranks

Since the results of the following sections shall not be discussed here, I restrict myself to some short remarks. We assume that Δ is a thick spherical building of irreducible type.

For Moufang buildings of rank 2 and for rank 3 buildings, the sphericity of $\Delta^0(c)$ is converted by Lemma 18 and by Lemma 19 into a group theoretic property. We are going to describe a similar translation for higher ranks now. Fix an apartment Σ of Δ such that $c \in \Sigma$, set $c^0 := \mathrm{op}_\Sigma(c)$, denote by Φ the set of all roots of Σ and by Φ_+ the subset of all those containing c. Recall that Δ is automatically Moufang if $n := \mathrm{rk}\,\Delta \geq 3$ (cf. [T3], Section 3). Denote by U_α ($\alpha \in \Phi$) the root group corresponding to α and set $U := \langle U_\alpha \mid \alpha \in \Phi_+ \rangle$ which is a subgroup of $\mathrm{Aut}(\Delta)$. Finally, we define $U_x := \mathrm{Stab}_U(x)$ for any vertex x of c^0. It follows from Lemma 17 ii) that the nerve of the covering of U by all cosets uU_x ($u \in U, x \in c^0$) is isomorphic to $\Delta^0(c)$. Hence we obtain:

Lemma 20: $\Delta^0(c)$ is $(n-1)$-spherical if and only if the family $\{U_x \mid x$ is a vertex of $c^0\}$ is $(n-1)$-generating for U in the sense of [AH].

□

Unfortunately, I do not know whether this criterion can be used efficiently for buildings of high rank in order to solve the sphericity problem for $\Delta^0(c)$. In [AH], the reverse direction is emphasized in similar situations, i.e. higher generation is deduced once certain homotopy properties are settled.

On the other side, the geometric/combinatorial approach to be discussed in the following sections cannot be applied to a number of finite buildings which is increasing exponentially with the rank. Nevertheless, only for the ground fields $\mathbb{F}_2$ and $\mathbb{F}_3$ explicit counter-examples are known if $n \geq 4$.

Example 9 (cf. [Bous], Chapter 2): Let $\Delta(X, \mathbb{F}_q)$ be the building associated to a Chevalley group of type $X \in \{A_n, C_n\}$ over $\mathbb{F}_q$, $\chi(X, q)$ the Euler characteristic of $\Delta^0(c)$ for $\Delta = \Delta(X, \mathbb{F}_q)$, $c \in \mathcal{C}(\Delta)$ and $\chi'(X, q) := (-1)^{n-1}(\chi(X, q) - 1)$. Then for $q \in \{2, 3\}$, $\chi'(X, q)$ is negative (and hence $\Delta^0(c)$ is not $(n-1)$-spherical) in "many" cases. For example, $\chi'(X, q) < 0$ for

$$X = A_n,\ q = 2 \quad \text{and} \quad n \in \{3, 4, 5, 6, 11, 12, 13, 18, \ldots\}$$
$$X = A_n,\ q = 3 \quad \text{and} \quad n \in \{10, 11, 12, \ldots, 21, 33, 34, \ldots\}$$
$$X = C_n,\ q = 2 \quad \text{and} \quad n \in \{3, 4, 5, 10, 11, 12, 13, 17, \ldots\}$$
$$X = C_n,\ q = 3 \quad \text{and} \quad n \in \{9, 10, 11, \ldots, 20, 32, 33, \ldots\}$$

the complete list of the signs of $\chi'(X, q)$ for $q \in \{2, 3\}$, $X \in \{A_n, C_n\}$ and $n \leq 100$ can be found in loc.cit.

The lack of further counter-examples is due to computational difficulties in my opinion since it is hardly possible to calculate the homology groups of $\Delta^0(c)$ directly if $\text{rk}\,\Delta \geq 4$ and the ground field contains at least 4 elements. However, I expect that the following is true: Given a finite field $\mathbb{F}_q$, there exists an $n = n(q)$ such that $\Delta^0(c)$ is not $(n-1)$-spherical for $\Delta = \Delta(A_n, \mathbb{F}_q)$.

I close this section by recalling a counter-example stated in [Ab1], §4, Remark 7. It shows that $\Delta^0(a)$ can also be non-spherical if $\text{rk}\,\Delta > 3$ and a is a vertex of Δ .

Example 10: Consider the building $\Delta = \Delta(A_4, \mathbb{F}_2)$ as the flag complex associated to the poset $\{0 < W < \mathbb{F}_2^5 \,|\, W \text{ is a proper subspace of } \mathbb{F}_2^5\}$. Let $a \in \Delta$ be a vertex corresponding to a subspace of dimension 2 or 3. Then $|\Delta^0(a)|$ is simply connected and $H_2(\Delta^0(a)) = \mathbb{Z}/2\,\mathbb{Z}$.

§ 3   General remarks concerning the sphericity proofs

We keep the notations introduced at the beginning of this chapter and shall only consider **thick** buildings henceforth. It is our goal to deduce sphericity properties of $\Delta^0(a)$ in the case of "classical" spherical buildings.

Definition 8: *A spherical building Δ is called **classical** if it satisfies one of the following three conditions:*

i) *Δ is the flag complex of a Desarguesian projective space.*

ii) *Δ is the flag complex of an embeddable polar space.*

iii) *Δ is a building of type D_n .*

Remark 10:

i) It is well known (cf. [T1], §6, or [Sch], §4.1) that every A_n building is the flag complex of an n-dimensional projective space and is therefore classical for $n \geq 3$.

ii) According to [T1], §7, every C_n building is the flag complex of a polar space of rank n and according to §8, every polar space of rank ≥ 3 whose maximal subspaces are Desarguesian is embeddable. In particular, every C_n building is classical for $n \geq 4$.

iii) It follows that every spherical building of irreducible type and rank ≥ 9 is classical.

iv) Every finite Moufang building of type A_n, C_n or D_n is also classical (for C_2 this follows from [FS]).

v) The classical buildings are precisely the buildings associated to the "natural" BN-pairs of classical groups (cf. for example [BrT1], §10).

The results of §2 point to some aspects which have to be taken into account while analysing the homotopy properties of $|\Delta^0(a)|$.

1. As Proposition 9 shows, it is not possible to deduce that $\Delta^0(a)$ is "usually" spherical by applying exclusively the abstract theory of spherical buildings as presented in [T1], §3. Therefore, the most natural and aesthetic proofs of the original Solomon–Tits theorem (cf. for example [Br3], §4.6) cannot be transferred to our situation.

2. If one adds the Moufang condition, still the exceptional behaviour of "small" buildings has to be taken care of. So the magnitude of the building (or to be more precise: the number of chambers containing a given panel) is an additional parameter which has to enter the reasoning somehow. Hence a second family of proofs of the Solomon–Tits theorem, depending on group theoretic arguments which do not admit any exceptions, cannot be modified appropriately for our purposes.

3. Even for rank 2 Moufang buildings, classification theorems have to be used at least partially in order to solve the sphericity problem for $\Delta^0(a)$. This indicates that one has to apply Tits' classification (cf. [T1]) in the case of spherical buildings of irreducible type and rank ≥ 3 .

Though I intensively looked for nicer proofs (if you work through §6 and §7 below, you will know why), I found only one approach respecting all three aspects and yielding the desired results at least for classical buildings. This approach is based on Quillen's proof of the Solomon–Tits theorem for buildings of type A_n

(cf. [Q]). It was already successfully used in [V], [Ab1] and [AA] in order to derive sphericity results for certain subcomplexes of A_n buildings. I am now going to describe the basic idea in terms which apply to all buildings of spherical type.

Let κ be a $(\dim \Delta)$-dimensional full subcomplex of Δ , e.g. $\kappa = \Delta^0(a)$, and let x_0 be a vertex of κ . Denote by κ_0 the subcomplex of κ consisting of all simplices $y \in \kappa$ such that no vertex of y is opposite to x_0 and the full convex hull of y and x_0 is contained in κ . Then $|\kappa_0|$ can be contracted along geodesics onto x_0 . (The set underlying $|\Delta|$ may be endowed with a metric inducing the usual metric on each sphere $|\Sigma|$ for any apartment Σ of Δ .) Now we try to contruct an increasing sequence $\kappa_0 \subset \kappa_1 \subset \ldots \subset \kappa_\ell = \kappa$ of subcomplexes of κ such that the following holds for all $1 \leq i \leq \ell$:

(1) If V_i denotes the set of vertices of κ_i not contained in κ_{i-1} , then
$$\kappa_i = \kappa_{i-1} \cup \bigcup_{x \in V_i} \mathrm{st}_{\kappa_i}(x) \text{ and } \mathrm{st}_{\kappa_i}(x) \cap \mathrm{st}_{\kappa_i}(x') \subseteq \kappa_{i-1} \text{ for all } x \neq x' \in V_i \ .$$

(2) $L_i(x) := \mathrm{st}_{\kappa_i}(x) \cap \kappa_{i-1}$ is $(n-2)$-spherical for any $x \in V_i$, where $n := \mathrm{rk}\,\Delta$.

Then suitable reduced Mayer–Vietoris sequences (cf. [Sp], Ch. 4, Sec. 6) yield $\widetilde{H}_j(\kappa_i) = \widetilde{H}_j(\kappa_{i-1}) = \ldots = \widetilde{H}_j(\kappa_0) = 0$ for all i and all $j \leq n - 2$. If $n \geq 3$, the theorem of Seifert–Van Kampen implies that all $|\kappa_i|$ are simply connected. Using the Hurewicz isomorphism (cf. [Sp], Ch. 7, Sec. 5), we obtain the following

Lemma 21: *If the conditions (1) and (2) are satisfied, then κ is $(n-1)$-spherical.*

□

This lemma shall be applied in the following way: Fix a class $\mathbf{C}$ of pairs (κ', Δ') , where Δ' is always a spherical building and κ' a $(\dim \Delta')$-dimensional full subcomplex of Δ' . In particular, we require $\kappa' \neq \emptyset$ for $\mathrm{rk}\,\Delta' = 1$, which is the basis of the induction below. Let κ and Δ be as above and suppose $(\kappa, \Delta) \in \mathbf{C}$. Assume that the κ_i are full subcomplexes of Δ and that all vertices in V_i are of the same type. Then condition (1) is satisfied automatically. If the κ_i can be chosen such that additionally $(L_i(x), L(x)) \in \mathbf{C}$ for any $x \in V_i$ and $L(x) := \ell k_\Delta(x)$, then (2) can be assumed by using induction on $n = \mathrm{rk}\,\Delta$. Thus we have proved the following

Lemma 22: *Let $\mathbf{C}$ be a class of pairs as described above. Assume that for any $(\kappa, \Delta) \in \mathbf{C}$ with $\mathrm{rk}\,\Delta \geq 2$, there exists a filtration $\kappa_0 \subset \kappa_1 \subset \ldots \subset \kappa_\ell = \kappa$ of κ such*

that the following is satisfied (set again $V_i := \{vertices\ of\ \kappa_i\} \setminus \{vertices\ of\ \kappa_{i-1}\}$):

(0) $|\kappa_0|$ *is contractible.*

(1)′ κ_i *is a full subcomplex of Δ for all $0 \leq i \leq \ell$, and all vertices in V_i are of the same type if $i \geq 1$.*

(2)′ $(\mathrm{st}_{\kappa_i}(x) \cap \kappa_{i-1}, \ell k_\Delta(x)) \in \mathbf{C}$ *for any $x \in V_i$ and $i \geq 1$.*

Then κ is $(\dim \Delta)$-spherical for any $(\kappa, \Delta) \in \mathbf{C}$. □

Lemma 22 is rather a program than a result. I mention some of the difficulties we shall have to overcome in the next section.

Problems:

i) The main difficulty consists in finding suitable classes **C** which are invariant under the induction step described in condition (2)′. Of course, we are first of all interested in pairs of the form $(\Delta^0(a), \Delta)$, where Δ is a spherical building (of given type) and $a \in \Delta$. But the class of all these pairs is definitely too small. It has to be enlarged suitably in order to permit applications of Lemma 22.

ii) It is hardly possible to describe the relative links $L_i(x) = \mathrm{st}_{\kappa_i}(x) \cap \kappa_{i-1}$ abstractly. One has to use coordinatizations of the spherical buildings in question. Since the class **C** should contain the pairs of the form $(\Delta^0(a), \Delta)$, we also need to know concrete descriptions of the abstractly defined subcomplexes $\Delta^0(a)$.

iii) If we start with a pair $(\kappa, \Delta) \in \mathbf{C}$ with $n = \mathrm{rk}\,\Delta$ and consider an inductively defined sequence of relative links $(L^{(0)} := \kappa,\ x_j \in L^{(j-1)})$

$$L^{(1)} = L_{i_1}(x_1) \subseteq \Delta^{(1)} := \ell k_\Delta(x_1), \ldots,$$
$$L^{(n-1)} = L_{i_{n-1}}(x_{n-1}) \subseteq \Delta^{(n-1)} := \ell k_{\Delta^{(n-2)}}(x_{n-1}),$$

we have to guarantee $L^{(n-1)} \neq \emptyset$. This leads to the requirement that every panel of Δ is contained in "sufficiently many" chambers and excludes for every n a (probably too big) finite number of finite buildings from our investigations.

iv) It was not necessary to specify the contractible subcomplex κ_0 of κ in the statement of Lemma 22. However, there is hardly any chance to satisfy $(2)'$ if κ_0 is "too small". Hence our candidate for κ_0 will be the complex described before Lemma 21 (possibly slightly reduced in order to obtain more symmetric relative links $L_i(x)$). Now κ_0 should be a full subcomplex of Δ which means the following here: Given simplices $a, b \in \kappa$ not containing any vertices opposite to x_0 such that $a \cup b \in \Delta$ and the convex hull of a and x_0 as well as of b and x_0 is included in κ, then the convex hull of $a \cup b$ and x_0 is also contained in κ. In the case of A_n and C_n buildings, this conclusion is in fact true if x_0 is chosen appropriately, meaning that x_0 corresponds to a point in the associated projective or polar space. But for D_n buildings, such a choice of x_0 is not possible, and this creates new problems.

I will close this section by briefly discussing a more special situation which is present in the cases A_n, C_n and F_4. Assume that we are considering spherical buildings which are flag complexes of certain geometries. If Γ is such a geometry of rank n, we denote by X_j $(1 \le j \le n)$ the set of its subspaces of type j, where type $=$ projective dimension $+ 1$ in the case of projective or polar spaces. Set $X := \bigcup_{j=1}^{n} X_j$, and write $x < x'$ if $x, x' \in X$ and x is a proper subspace of x'. We assume that $x < x'$ implies type $(x) <$ type (x'). Now $\Delta = \operatorname{Flag} X$ is the simplicial complex associated to the poset X, i.e. every d-dimensional simplex of Δ is a chain of the form $x_1 < \ldots < x_{d+1}$ with $x_k \in X$ for all $1 \le k \le d+1$. Note that $1 \le$ type $(x_1) < \ldots <$ type $(x_{d+1}) \le n$. For any vertex $x \in X$, we define $X^{<x} := \{x' \in X \mid x' < x\}$, $X^{>x} := \{x' \in X \mid x < x'\}$, $\Delta^{<x} := \operatorname{Flag} X^{<x}$, $\Delta^{>x} = \operatorname{Flag} X^{>x}$ and obtain $\ell k_\Delta(x) = \Delta^{<x} * \Delta^{>x}$.

If κ is a full subcomplex of Δ and $Y \subseteq X$ the set of its vertices, then $\kappa = \operatorname{Flag} Y$. To filter κ by full subcomplexes $\kappa_0 \subset \kappa_1 \subset \ldots \subset \kappa_\ell = \kappa$ means to filter Y by subsets $Y_0 \subset \ldots \subset Y_\ell = Y$ and to set $\kappa_i = \operatorname{Flag} Y_i$. We get $V_i = Y_i \setminus Y_{i-1}$, and $(1)'$ is satisfied if $V_i \subseteq X_j$ for some $j = j(i)$ $(1 \le i \le \ell)$. Furthermore,

$$L_i(x) = \ell k_\Delta(x) \cap \kappa_{i-1} = (\operatorname{Flag} Y_{i-1}^{<x}) * (\operatorname{Flag} Y_{i-1}^{>x}) =: \kappa_{i-1}^{<x} * \kappa_{i-1}^{>x}$$

for any $x \in V_i$. It is now sufficient to investigate the complexes $\kappa_{i-1}^{<x}$ and $\kappa_{i-1}^{>x}$ separately since the sphericity of a join follows from the sphericity of the two factors. Alternatively, we may assume that the class $\mathbf{C}$ in question is closed under formation of joins. Then it suffices to check

$(\kappa_{i-1}^{\leq x}, \Delta^{<x}) \in C$ and $(\kappa_{i-1}^{\geq x}, \Delta^{>x}) \in C$ $(x \in V_i,\ 1 \leq i \leq \ell)$ in order to verify condition (2)$'$.

So if we are dealing with flag complexes of geometries, the program we have to work off is the following:

- Describe $\Delta^0(a)$ explicitly as a subcomplex of Flag $X = \Delta$.

- Find a suitable class C containing all pairs $(\Delta^0(a), \Delta)$ for "sufficiently big" Δ .

- Given $(\kappa = \mathrm{Flag}\, Y, \Delta = \mathrm{Flag}\, X) \in C$, construct a filtration $Y_0 \subset \ldots \subset Y_\ell = Y$ such that $|\,\mathrm{Flag}\, Y_0|$ is contractible and $V_i = Y_i \setminus Y_{i-1}$ consists of vertices of the same type if $i \geq 1$.

- Verify $(\kappa_{i-1}^{\leq x}, \Delta^{<x})$, $(\kappa_{i-1}^{\geq x}, \Delta^{>x}) \in C$ for all $x \in V_i,\ 1 \leq i \leq \ell$.

Of course, during the search for a complete proof, the last three points cannot be treated linearly one after the other. They have to be attacked simultaneously since the intended filtrations influence the choice of the class C .

So far for the general principles. Let us look at the details now.

§ 4   The case A_n

The general remarks of the preceding section shall first be illustrated by means of the simplest case, namely A_n buildings. The conclusions as well as parts of the argumentation will also be used while dealing with the more difficult cases C_n and D_n . The main result of this section, namely Proposition 12 below, is not new but contained in [AA]. However, in order to present a self-contained treatment of all classical buildings, those of type A_n shall briefly be discussed here as well.

Let K be a skew field and V an $(n+1)$-dimensional vector space over K. Set $X := X(V) := \{0 < U < V \,|\, U \text{ is a } K\text{-subspace of } V\}$ and $\Delta = \mathrm{Flag}\, X$. This is the standard description of a classical A_n building. For any $M \subseteq V$, we denote by $\langle M \rangle$ the K-subspace generated by M . If $e_1, \ldots, e_{n+1}$ is a basis of V ,
$$\Sigma(e_1, \ldots, e_{n+1}) := \mathrm{Flag}\{\langle e_{i_1}, \ldots, e_{i_r}\rangle \,|\, 1 \leq i_1 < \ldots < i_r \leq n+1, 1 \leq r \leq n\}$$
is an apartment of Δ . Every apartment Σ of Δ is of the form $\Sigma = \Sigma(e_1, \ldots, e_{n+1})$ for some basis. It is well known (cf. [Ab4], Lemma 1.2.1, if necessary) that $\mathrm{op}_\Sigma(\langle e_{i_1}, \ldots, e_{i_r}\rangle) = \langle e_{j_1}, \ldots, e_{j_{n+1-r}}\rangle$, where $\{j_1, \ldots, j_{n+1-r}\} = \{1, \ldots, n+1\} \setminus \{i_1, \ldots, i_r\}$. From this it follows

Lemma 23: *Two vertices* $U, W \in X$ *are opposite in* Δ *if an only if* $V = U \oplus W$. *Hence* $\Delta^0(U) = \mathrm{Flag}\{W' \in X \mid W' \cap U = 0 \text{ or } W' + U = V\}$.

$\square$

This observation leads to the following definition which is important for C_n and D_n as well:

Definition 9:

i) *Two subspaces* U *and* W *of* V *are called* **transversal** *(in* V *) if* $U \cap W = 0$ *or* $U + W = V$. *If this is the case, we write* $U \pitchfork_V W$ *or simply* $U \pitchfork W$.

ii) *For a given set* $\mathcal{E}$ *of subspaces of* V, $U \pitchfork \mathcal{E}$ *means by definition that* $U \pitchfork E$ *for all* $E \in \mathcal{E}$. *We set*

$$X_{\mathcal{E}}(V) := \{U \in X \mid U \pitchfork \mathcal{E}\} \quad \text{and} \quad T_{\mathcal{E}}(V) := \mathrm{Flag}\, X_{\mathcal{E}}(V) .$$

In view of Lemma 16 iii), Lemma 23 implies

Corollary 12: *For any simplex* $a = \{E_1 < \ldots < E_r\} \in \Delta$ *and* $\mathcal{E}(a) := \{E_i \mid 1 \le i \le r\}$, $\Delta^0(a) = \mathrm{Flag}\{U \in X \mid U \pitchfork \mathcal{E}(a)\} = T_{\mathcal{E}(a)}(V)$. $\square$

In order to carry out the planned induction, we shall have to consider all subcomplexes of the form $T_{\mathcal{E}}(V)$, where $\mathcal{E}$ is a finite set of subspaces of V . Before doing this, we recall two further standard facts concerning A_n buildings:

- For any d-dimensional subspace U of V , $\Delta^{<U}$ is the A_{d-1} building associated to $X^{<U}$ and $\Delta^{>U}$ is the A_{n-d-1} building associated to $X(V/U)$.

- Let ℓ be a line in V , i.e. $\dim \ell = 1$, and $a = \{U_1 < \ldots < U_r\} \in \Delta$ a simplex not containing any vertex opposite to ℓ . Then the full convex hull of ℓ and a in Δ is the flag complex of $\{U_i, U_i + \ell, \ell \mid 1 \le i \le r\}$.

Proposition 12 (cf. [AA], Theorem 1.1): *Let* $\mathcal{E}$ *be a finite set of subspaces of* V , $\mathcal{E}_j := \{E \in \mathcal{E} \mid \dim E = j\}$ *and* $e_j := \#\mathcal{E}_j$. *Assume that* $\#K \ge \sum_{j=1}^{n} \binom{n-1}{j-1} e_j$. *Then* $T_{\mathcal{E}}(V)$ *is* $(n-1)$-*spherical.*

Proof: We consider the class **C** of all pairs $(T_{\mathcal{E}'}(V'), \text{Flag } X')$ with $\sum_{j=1}^{n'} \binom{n'-1}{j-1} e'_j \leq \#K'$ and joins of these pairs. One has to show that $\kappa = T_{\mathcal{E}}(V)$ admits a filtration as described in Lemma 22 (cf. also the discussion at the end of §3).

1. There exists a line $\ell \in X_{\mathcal{E}}(V) =: Y$

This is clear if K is infinite. If $q := \#K < \infty$, the union $\bigcup_{E \in \mathcal{E}} E$ contains at most

$\frac{q^n-1}{q-1} \sum_{j=1}^{n} e_j \leq q \frac{q^n-1}{q-1} < \frac{q^{n+1}-1}{q-1}$ lines. Hence there exists a line ℓ in V with $\ell \pitchfork \mathcal{E}$.

2. Description of the filtration

Choose a line $\ell \in Y$ and set $Y_0 := \{U \in Y \mid U + \ell \in Y\}$. Then $|\text{Flag } Y_0|$ can be contracted via $U \longmapsto U + \ell \longmapsto \ell$ onto ℓ. (This follows even without considering convex hulls in Δ.) If U does not contain ℓ, we observe that $\dim(U+\ell) \cap E = \dim U \cap (E+\ell)$ for any $E \in \mathcal{E} \setminus \{V\}$ since $\ell \pitchfork \mathcal{E}$. Hence $U + \ell \pitchfork \mathcal{E}$ is equivalent to $U \pitchfork \mathcal{E} + \ell$, where of course $\mathcal{E} + \ell := \{E + \ell \mid E \in \mathcal{E}\}$. This implies $Y_0 = \{U \in X \mid (\ell \leq U \text{ and } U \pitchfork \mathcal{E}) \quad \text{or} \quad (\ell \not\leq U \text{ and } U \pitchfork \mathcal{E} \cup (\mathcal{E} + \ell))\}$.

For any integer $0 \leq i \leq n$, we set $Y_i := Y_0 \cup \{U \in Y \mid \dim U \geq n+1-i\}$ and $\kappa_i := \text{Flag } Y_i$. In particular, $\kappa_n = \kappa = T_{\mathcal{E}}(V)$ and $Y_i \setminus Y_{i-1}$ $(1 \leq i \leq n)$ contains only elements of type $n + 1 - i$, i.e. only $(n+1-i)$-dimensional subspaces of V.

3. Determination of the relative links

Fix an i with $1 \leq i \leq n$ and a $U \in Y_i \setminus Y_{i-1}$. In particular, $\dim U = n+1-i$, $U \pitchfork \mathcal{E}$ and $\ell \not\leq U$. We have to consider $Y_{i-1}^{<U} = Y_0^{<U}$ and $Y_{i-1}^{>U} = Y^{>U}$.

In view of $U \pitchfork \mathcal{E}$, one obtains for all $W < U$:

$$W \underset{V}{\pitchfork} \mathcal{E} \iff W \underset{U}{\pitchfork} \mathcal{E} \cap U := \{E \cap U \mid E \in \mathcal{E}\}$$

One just has to observe that $W \cap E = W \cap (E \cap U)$ and that $W + E = V \iff W + E \supseteq U \iff W + (E \cap U) = U$ in case $U + E = V$.

Furthermore, $W < U$ and $U \pitchfork E$ imply $W \underset{V}{\pitchfork} E + \ell \iff W \underset{U}{\pitchfork} (E + \ell) \cap U$. As above, this is clear if $U \pitchfork E + \ell$. So let us assume $U + E + \ell \neq V$ and $U \cap (E + \ell) \neq 0$. Then it follows $U + E \neq V$, $U \cap E = 0$ and $\dim U \cap (E + \ell) = 1$. This implies

$$W \mathbin{\underset{V}{\pitchfork}} E + \ell \iff W \cap (E + \ell) = 0 \iff W \cap (E + \ell) \cap U = 0$$
$$\iff W \mathbin{\underset{U}{\pitchfork}} (E + \ell) \cap U$$

for any $W < U$. Therefore, all $W < U$ satisfy the equivalence
$$W \mathbin{\underset{V}{\pitchfork}} \mathcal{E} + \ell \iff W \mathbin{\underset{U}{\pitchfork}} (\mathcal{E} + \ell) \cap U .$$

Setting $\mathcal{E}' := (\mathcal{E} \cap U) \cup ((\mathcal{E} + \ell) \cap U)$, we obtain
$Y_0^{<U} = \{0 < W < U \mid W \mathbin{\underset{U}{\pitchfork}} \mathcal{E}'\} = X_{\mathcal{E}'}(U)$. Now observe that
$e_1' := \# \mathcal{E}_1' \le e_1 + \ldots + e_{i+1}$ and $e_j' := \# \mathcal{E}_j' \le e_{j+i-1} + e_{j+i}$ for $2 \le j \le n-i$. Therefore
$$\sum_{j=1}^{n-i} \binom{n-i-1}{j-1} e_j' \le e_1 + \ldots + e_{i-1} + \sum_{r=i}^{n} e_r \left(\binom{n-i-1}{r-i} + \binom{n-i-1}{r-i-1} \right) \le$$
$$\le e_1 + \ldots + e_{i-1} + \sum_{r=i}^{n} \binom{n-1}{r-i} e_r \le \sum_{j=1}^{n} \binom{n-1}{j-1} e_j \le \#K$$

This shows that $(\kappa_{i-1}^{<U}, \Delta^{<U}) = (\operatorname{Flag} X_{\mathcal{E}'}(U), \operatorname{Flag} X(U)) \in \mathbf{C}$.

In order to describe $Y^{>U}$, we set $\overline{V} := V/U$, $\overline{W} := W/U$ for any $W > U$,
$\overline{E} := E + U/U$ and $\overline{\mathcal{E}} := \{\overline{E} \mid E \in \mathcal{E}\}$. Since $U \mathbin{\pitchfork} \mathcal{E}$, it is easy to verify
$W \mathbin{\underset{V}{\pitchfork}} \mathcal{E} \iff \overline{W} \mathbin{\underset{\overline{V}}{\pitchfork}} \overline{\mathcal{E}}$ for all $W > U$. Therefore the poset

$\{U < W < V \mid W \mathbin{\pitchfork} \mathcal{E}\} = Y^{>U} = Y_{i-1}^{\ge U}$ is isomorphic to
$\{0 < \overline{W} < \overline{V} \mid \overline{W} \mathbin{\underset{\overline{V}}{\pitchfork}} \overline{\mathcal{E}}\} = X_{\overline{\mathcal{E}}}(\overline{V})$. Notice also that

$$\sum_{j=1}^{i-1} \binom{i-2}{j-1} \overline{e}_j \le \sum_{j=1}^{i-1} \binom{i-2}{j-1} e_j \le \sum_{j=1}^{n} \binom{n-1}{j-1} e_j \le \#K$$

Hence $(\kappa_{i-1}^{\ge U}, \Delta^{>U}) \cong (\operatorname{Flag} X_{\overline{\mathcal{E}}}(\overline{V}), \operatorname{Flag} X(\overline{V})) \in \mathbf{C}$. $\qquad\square$

Corollary 13: *If* $a = \{E_1 < \ldots < E_r\} \in \Delta$ *and* $\dim E_i =: d_i \ (1 \le i \le r)$, *then*
$\Delta^0(a)$ *is* $(n-1)$-*spherical provided that* $\#K \ge \sum_{i=1}^{r} \binom{n-1}{d_i - 1}$. *In particular,* Δ *possesses*
the property (S_Δ) *introduced in Chapter I, §5, Remark 6, if* $\#K \ge 2^{n-1}$. $\qquad\square$

Remark 11:

i) For $\mathcal{E} = \emptyset$ and $T_{\mathcal{E}}(V) = \Delta$, one recovers Quillen's original proof of the Solomon–
Tits theorem for A_n buildings.

ii) The case where $\mathcal{E} = \{E_1, E_2\}$, $\dim E_1 = 1$ and $\dim E_2 = n$, was first treated
(for arbitrary K) by Vogtmann in [V], Proposition 1.4.

iii) For $\mathcal{E}$ consisting of a single d-dimensional subspace E of V and $\#K \geq \binom{n-1}{d-1}$, the sphericity of $T_{\mathcal{E}}(V)$ was proved in [Ab1], §4.

iv) Instead of using a line $\ell \in X_{\mathcal{E}}(V)$ in order to define Y_0 , one can dually work with a hyperplane $H \in X_{\mathcal{E}}(V)$ and then filter dimension – increasingly. In fact this is done in [AA].

v) The occurrence of the binomial coefficents $\binom{n-1}{j-1}$ in Proposition 12 is due to the estimation $e'_j \leq e_{j+i-1} + e_{j+i}$ in the proof above. For C_n buildings, one obtains $\bar{e}_j \leq e_{j+i-1} + 2e_{j+i} + e_{j+i+1}$ while analyzing $Y_0^{>U}$. This leads to binomial coefficients of the form $\binom{2(n-1)}{j-1}$ in Proposition 13 of §6.

§ 5   Lemmata on hermitian and pseudo–quadratic forms

Throughout the next two sections, we shall work with the coordinatizations of classical C_n and of D_n buildings developed in [T1], §8. Roughly speaking, these buildings belong to vector spaces endowed with a hermitian or pseudo-quadratic form. Some facts concerning these forms will be established in the present preparatory section. Though I did not find them in the literature, I do not claim that all these results, elementary as they are, appear here for the first time. For further information concerning hermitian forms, the reader is referred to [HO] and [Bou1]. Throughout the rest of this chapter, we shall use the following

Notations and agreements:

K is a skew field

$\sigma : K \longrightarrow K$, $\alpha \longmapsto \alpha^\sigma$, is an involution, i.e. an anti-automorphism of K satisfying $\sigma^2 = \mathrm{id}_K$

$\varepsilon \in \{1, -1\} \subset K$; $\varepsilon = -1$ if $\sigma \neq \mathrm{id}_K$
$K_{\sigma,\varepsilon} := \{\alpha - \alpha^\sigma \varepsilon \,|\, \alpha \in K\}$,
$K^{\sigma,\varepsilon} := \{\alpha \in K \,|\, \alpha + \alpha^\sigma \varepsilon = 0\}$

Λ is a form parameter relative to (σ, ε) , i.e. Λ is a subgroup of $(K, +)$ satisfying $K_{\sigma,\varepsilon} \subseteq \Lambda \subseteq K^{\sigma,\varepsilon}$ and $\alpha^\sigma \Lambda \alpha \subseteq \Lambda$ for all $\alpha \in K$.

V is a right K-vector space of dimension $m \in \mathbb{N} \cup \{\infty\}$

$f : V \times V \longrightarrow K$ is a (σ, ε)-**hermitian form**, i.e. f is biadditive and

$f(x\alpha, y\beta) = \alpha^\sigma f(x,y)\beta$, $f(y,x) = f(x,y)^\sigma \varepsilon$ for all $x, y \in V$, $\alpha, \beta \in K$

$Q : V \longrightarrow K/\Lambda$ is a (σ, ε)-**quadratic form** with associated (σ, ε)-hermitian form f, i.e. $Q(x\alpha) = \alpha^\sigma Q(x)\alpha + \Lambda$ and $Q(x + y) - Q(x) - Q(y) = f(x,y) + \Lambda$ for all $x, y \in V$, $\alpha \in K$.

If $\Lambda \neq K$, f is uniquely determined by this last equation and automatically trace-valued, i.e. $f(x,x) \in \{\alpha + \alpha^\sigma \varepsilon \,|\, \alpha \in K\} \; \forall x \in V$. If $\Lambda = K$, we require that f is alternating and hence also trace-valued in this case.

For any subset M of V , we set $M^\perp := \{x \in V \,|\, f(x, M) = 0\}$.

A subspace U of V is called **non-degenerate** if $U \cap U^\perp = 0$, **totally degenerate** if $U \subseteq U^\perp$, **anisotropic** if $0 \notin Q(U \setminus \{0\})$, **isotropic** if $0 \in Q(U \setminus \{0\})$ and **totally isotropic** if $U \subseteq U^\perp$ and $Q(U) = 0$. Denote by n the **Witt index** of (V, Q, f) , i.e. the common dimension of all maximal totally isotropic subspaces of V . We require that $0 < n < \infty$.

Remark 12:

i) The terminology introduced above agrees extensively with that of [T1], §8, but the notions "non-degenerate", "totally degenerate" and "totally isotropic" are used slightly different here, namely in the sense of [HO]. Note also that pseudo-quadratic forms are only defined relative to the form parameter $\Lambda = K_{\sigma,\varepsilon}$ in [T1]. Since we are a little more flexible at this point, we can assume that the quadratic space V parametrizing a classical C_n building is always non-degenerate "in the narrow sense", i.e. $V^\perp = 0$ (cf. Lemma 27 below).

ii) If $\Lambda = K^{\sigma,\varepsilon}$, then a subspace of V is totally isotropic if and only if it is totally degenerate.

iii) If the characteristic of K is $\neq 2$ or if σ is not the identity on the center of K, then $K_{\sigma,\varepsilon} = K^{\sigma,\varepsilon}$ and Q is uniquely determined by f . Hence it suffices to consider just (V, f) instead of (V, Q, f) in these cases.

We are going to show now that V contains "enough" isotropic lines if K is infinite. As a first step, we prove

Lemma 24: *Assume that K is infinite and $(\sigma, \varepsilon) \neq (\mathrm{id}, 1)$. Then $K_{\sigma,\varepsilon}$ is also infinite.*

Proof: According to [HO], Lemma 6.1.1, the fixed field $k := K^{\sigma,-1}$ of σ is infinite. In view of Remark 12 iii), this already establishes the assertion in most cases. So let us assume char $K = 2$, $\varepsilon = 1$ and $\sigma \neq \mathrm{id}$ now. Suppose that $K_{\sigma,1}$ is finite. Then it follows inductively that $\bigcap\limits_{i=1}^{r} C_k(s_i)$ is infinite for any finite subset $\{s_1, \ldots, s_r\}$ of k, where $C_k(s_i) := \{\alpha \in k \,|\, \alpha s_i = s_i \alpha\}$. One just has to observe that $\bigcap\limits_{i=1}^{r} C_k(s_i)$ is the kernel of the additive map

$$\varphi_r : \bigcap_{i=1}^{r-1} C_k(s_i) \longrightarrow K_{\sigma,1}, \quad \alpha \longmapsto \alpha s_r + s_r \alpha = \alpha s_r + (\alpha s_r)^\sigma$$

Now choose $\alpha_0 \in K_{\sigma,1} \setminus \{0\}$, define $s_1 := \alpha_0$, choose inducitively $s_j \in (\bigcap\limits_{i=1}^{j-1} C_k(s_i)) \setminus \{s_1, \ldots, s_{j-1}\}$ and set $\alpha_j := s_j^2 \alpha_0 = s_j^\sigma \alpha_0 s_j \in K_{\sigma,1}$ ($j \in \mathbb{N}$). Using $s_i s_j = s_j s_i$, one obtains

$$\alpha_i = \alpha_j \Rightarrow s_i^2 = s_j^2 \Rightarrow (s_i + s_j)^2 = 0 \Rightarrow s_i = s_j \Rightarrow i = j$$

Hence $\{\alpha_j \,|\, j \in \mathbb{N}\}$ is an infinite subset of $K_{\sigma,1}$, contrary to the assumption that $K_{\sigma,1}$ is finite which is therefore false. $\qquad\square$

Lemma 25: *Let K be infinite, V isotropic and $E_1, \ldots, E_r$ ($r \in \mathbb{N}$) proper subspaces of V. Assume that $V^\perp \cap Q^{-1}(0) = 0$ and that $m = \dim V \geq 3$ if $\Lambda = 0$. Then the set I of all isotropic vectors in V is not contained in $E := \bigcup\limits_{i=1}^{r} E_i$.*

Proof: We use induction on r, the basis being the well known fact that V is spanned by I (cf. [T1], §8, Lemma 8.2.7). Two cases have to be distinguished:

1. $\Lambda \neq 0$

   Choose $x \in I \setminus E_1$ and $y \in I \setminus \bigcup\limits_{i=2}^{r} E_i$. Then $U := xK + yK$ is not contained in E_i for $1 \leq i \leq r$. If $\dim U = 1$, we are ready. If U is a degenerate plane, it has to be totally isotropic and we may choose any vector in U outside $\bigcup\limits_{i=1}^{r} (U \cap E_i)$. Finally, let U be a hyperbolic plane. It contains infinitely many isotropic lines since $\#\Lambda = \infty$ by Lemma 24 (if $(\sigma, \varepsilon) = (\mathrm{id}, 1)$, use $\#K^2 = \infty$). Therefore again $U \cap I \not\subseteq \bigcup\limits_{i=1}^{r} (U \cap E_i)$.

79

2. $\Lambda = 0$

We first assume $m = 3$. Then it is easily checked by using a Witt basis of V that V contains infinitely many isotropic lines. On the other hand, every proper subspace of V contains at most 2 isotropic lines.

If $m > 3$, we choose $x \in I \setminus E_1$, $y \in I \setminus (E_3 \cup \ldots \cup E_r \cup \{x\}^\perp)$ and set $H := xK + yK$ which is a hyperbolic plane. Also in case $m = \infty$, it is easy to check that $H \cap H^\perp = 0$ and $H + H^\perp = V$. $H^\perp$ cannot be totally isotropic since it would be contained in $V^\perp \cap Q^{-1}(0)$ then. It is therefore possible to choose an anisotropic vector $z \in H^\perp$, and we may additionally assume $z \notin E_2$ if $H^\perp \not\subseteq E_2$. Set $U := H + zK$ and observe that $U \not\subseteq E_i$ for all $1 \leq i \leq r$ as well as $U^\perp \cap Q^{-1}(0) = 0$. Hence the proof is finished by applying the case $m = 3$ treated above. $\qquad\square$

Next we consider a finite field $K = \mathbb{F}_q$ with q elements and calculate the number of isotropic lines in V . Set $V_0 := V^\perp \cap Q^{-1}(0)$ and $m_0 := \dim V_0$. Choose a complement V_1 of V_0 in V , i.e. $V = V_0 \oplus V_1$, let m_1 be its dimension and n_1 its Witt index. Note that $m = m_0 + m_1$, $n = m_0 + n_1$ and $2n_1 \leq m_1 \leq 2n_1 + 2$. We define $\eta \in \{0, \frac{1}{2}, 1\}$ by $\#\Lambda =: q^{1-\eta}$, and set again $I := Q^{-1}(0) \setminus \{0\}$. Then the following holds:

Lemma 26: $\#\{\langle x \rangle = x\,\mathbb{F}_q \mid x \in I\} = \frac{1}{q-1}(q^{m-\eta} + q^n - q^{m-n_1-\eta} - 1)$

Proof: Choose n_1 pairwise orthogonal hyperbolic planes $H_1, \ldots, H_{n_1}$ in V_1 and isotropic vectors $e_i, e_{-i} \in H_i$ satisfying $f(e_i, e_{-i}) = 1$ $(1 \leq i \leq n_1)$. Let A be the (anisotropic) orthogonal complement of $\overset{n_1}{\underset{i=1}{\bot}} H_i$ in V_1 . Then any $x \in V$ can uniquely be written in the form

$$x = x_0 + \sum_{i=1}^{n_1}(e_i \lambda_i + e_{-i}\mu_i) + a \text{ with } x_0 \in V_0, \, \lambda_i, \mu_i \in \mathbb{F}_q \text{ and } a \in A .$$

Note that $Q(x) = Q(a) + (\sum_{i=1}^{n_1} \lambda_i^q \mu_i + \Lambda)$.

Now we define $I_j := \{x \in I \mid \lambda_1 = \ldots = \lambda_{j-1} = 0 \text{ and } \lambda_j = 1\}$ $(1 \leq j \leq n_1)$. Observe that $\#I_j = q^{m-j-1} \cdot \#\Lambda = q^{m-j-\eta}$. Since

$$\#\{\langle x \rangle \mid x \in I\} = \sum_{j=1}^{n_1} \#I_j + \#\{\langle x \rangle \mid x \in V_0 + \sum_{i=1}^{n_1} e_{-i}\,\mathbb{F}_q\} \text{ , the claim follows.} \qquad\square$$

Using Lemma 26 in order to estimate the number of isotropic lines in $E = \bigcup_{i=1}^{r} E_i$, one obtains the following analogue of Lemma 25 for finite fields:

Corollary 14: *Keep the notations introduced above. Assume that V is isotropic and $V_0 = 0$. Let $E_1, \ldots, E_r$ be proper subspaces of V. Then I is not contained in $E := \bigcup_{i=1}^{r} E_i$ provided that one of the following conditions holds:*

a) $\Lambda = \mathbb{F}_q$ *and* $q \geq r$

b) $\sigma \neq \mathrm{id}$, $m = 2$ *and* $q^{\frac{1}{2}} \geq r$

c) $m \geq 3$ *and* $q \geq 2r$

d) $m = 2n + 2$ $(\Rightarrow \Lambda = 0)$ *and* $q \geq r + 1$ $\qquad\qquad\qquad$ $\square$

We are interested in hermitian and pseudo-quadratic forms here because of the following two results of Tits, the first being rather elementary (cf. [T1], 8.3.4 and 8.4.2) and the second a deep theroem (cf. [T1], Theorem 8.22):

1. Let (V, Q, f) be as described at the beginning of this section, assume that $V_0 := V^{\perp} \cap Q^{-1}(0) = 0$ and set $X := \{0 < U < V \,|\, U$ is a totally isotropic subspace of $V\}$. Then Flag X is a C_n building which is thick unless $(m, \Lambda) = (2n, 0)$.

2. Every thick C_n building whose links of type A_2 correspond to Desarguesian planes (in particular, any C_n building for $n \geq 4$) is of the form Flag X as above.

Following an idea described in [HO], §5.2.B, in the context of unitary groups, one can achieve $V^{\perp} = 0$ in 1. by adjusting the choice of Λ. In order to see this, we can assume char $K = 2$ in view of Remark 12 ii) and iii). Define $\tilde{\Lambda} \supseteq \Lambda$ by $\tilde{\Lambda}/\Lambda := Q(V^{\perp})$. Then $\tilde{\Lambda} \subseteq K^{\sigma,\varepsilon}$ since $f(x,x) \equiv \alpha + \alpha^{\sigma}\varepsilon \bmod \Lambda$ for any $\alpha \in Q(x)$ (cf. [HO], 5.1.14). Hence $\tilde{\Lambda}$ is a form parameter. Set $\tilde{V} := V/V^{\perp}$, denote by $\pi : V \twoheadrightarrow \tilde{V}$ the canonical projection and by $\tilde{f}$ the (σ, ε)-hermitian form on $\tilde{V} \times \tilde{V}$ induced by f. Obviously, $\tilde{V}^{\perp} = 0$. Define $\tilde{Q} : \tilde{V} \longrightarrow K/\tilde{\Lambda}$ by $\tilde{Q}(x + V^{\perp}) := Q(x) + \tilde{\Lambda}$, and set $\tilde{X} := \{0 < \tilde{U} < \tilde{V} \,|\, \tilde{U}$ is totally isotropic relative to $(\tilde{Q}, \tilde{f})\}$.

Lemma 27:

i) *π induces a bijection between $Q^{-1}(0)$ and $\tilde{Q}^{-1}(0)$.*

ii) *The posets X and $\tilde{X}$ (and hence also their flag complexes) are isomorphic.*

Proof:

i) $\pi_{|Q^{-1}(0)}$ is injective since $V^{\perp} \cap Q^{-1}(0) = 0$. If $\tilde{Q}(x + V^{\perp}) = 0$, then

$Q(x) = Q(z)$ for some $z \in V^{\perp}$. Therefore, since char $K = 2$,

$Q(x + z) = Q(x) + Q(z) = 0$, i.e. $x + z \in \pi^{-1}(x + V^{\perp}) \cap Q^{-1}(0)$.

ii) Denote by $\kappa : \tilde{Q}^{-1}(0) \longrightarrow Q^{-1}(0)$ the inverse of $\pi_{|Q^{-1}(0)} : Q^{-1}(0) \longrightarrow \tilde{Q}^{-1}(0)$.

The restriction of κ to any totally isotropic subspace $\tilde{U}$ of $\tilde{V}$ is linear. In fact, given $\tilde{x}, \tilde{y} \in \tilde{U}$, $f(\kappa(\tilde{x}), \kappa(\tilde{y})) = \tilde{f}(\tilde{x}, \tilde{y}) = 0$ implies $Q(\kappa(\tilde{x}) + \kappa(\tilde{y})) = Q(\kappa(\tilde{x})) + Q(\kappa(\tilde{y})) = 0$ and hence $\kappa(\tilde{x} + \tilde{y}) = \kappa(\tilde{x}) + \kappa(\tilde{y})$. Therefore, $\kappa(\tilde{U})$ is a totally isotropic subspace of V. Now it is obvious that π and κ induce inclusion-preserving maps between X and $\tilde{X}$ which are inverse to each other. $\qquad\Box$

Remark 13: It is shown in [HO] that the unitary groups of (V, Q, f) and $(\tilde{V}, \tilde{Q}, \tilde{f})$ are isomorphic. The most prominent example is provided by the isomorphism between $O_{2n+1}(K)$ and $Sp_{2n}(K)$ for a perfect field K of characteristic 2.

So we may and will assume $V^{\perp} = 0$ in the following. But we have to admit infinite dimensional quadratic spaces in order to describe all classical C_n buildings. The question occurs wether the usual laws for taking orthogonal complements are still valid then. For our purposes, the following statements are sufficient.

Lemma 28: Set $\mathcal{U} := \{0 \le U \le V \mid \dim U < \infty\}$, $\mathcal{U}^{\perp} := \{U^{\perp} \mid U \in \mathcal{U}\}$ and $\mathcal{W} := \mathcal{U} \cup \mathcal{U}^{\perp}$. Then for any $A, B \in \mathcal{W}$, it holds:

i) $A^{\perp} \in \mathcal{W}$, $A^{\perp\perp} = A$ and $\mathrm{codim}_V A^{\perp} = \dim A$

ii) $(A + B)^{\perp} = A^{\perp} \cap B^{\perp}$

iii) $(A \cap B)^{\perp} = A^{\perp} + B^{\perp}$

iv) $A + B \in \mathcal{W}$ and $A \cap B \in \mathcal{W}$

v) $A \pitchfork B$ if and only if $A^{\perp} \pitchfork B^{\perp}$

82

Proof:

i) For any $U \in \mathcal{U}$, $U^{\perp\perp} = U$ and $\operatorname{codim}_V U^\perp = \dim U$ by Corollary 1 of Proposition 4 in [Bou1], §1, n° 6. This implies the non-trivial assertions in i).

ii) is obvious for arbitrary subspaces A, B of V.

iii) follows from Corollary 2 of Proposition 4 in loc.cit. if $A, B \in \mathcal{U}$. For $A \in \mathcal{U}$ and $B \in \mathcal{U}^\perp$, this equation is (almost) proved in loc.cit., §1, Exercise 9. If $A, B \in \mathcal{U}^\perp$, one obtains by applying i) and ii)
$$A^\perp + B^\perp = (A^\perp + B^\perp)^{\perp\perp} = (A^{\perp\perp} \cap B^{\perp\perp})^\perp = (A \cap B)^\perp .$$

iv) is an immediate consequence of the first three statements.

v) Applying i) – iv) yields
$$A \cap B = 0 \iff (A \cap B)^\perp = 0^\perp \iff A^\perp + B^\perp = V \text{ and}$$
$$A + B = V \iff (A + B)^\perp = V^\perp \iff A^\perp \cap B^\perp = 0$$

$\square$

§ 6 The case C_n

Concerning (V, Q, f), we keep the notations introduced at the beginning of §5. Recall that we always assume $V^\perp = 0$ now. Furthermore, we require $(m, \Lambda) \neq (2n, 0)$ throughout this section (ordinary quadratic spaces of dimension $2n$ and Witt index n will be considered in §7 in connection with D_n buildings). As already mentioned, one obtains a thick C_n building $\Delta = \operatorname{Flag} X$ by setting $X := X(V) := \{0 < U < V \mid U$ is totally isotropic$\}$, and every classical C_n building can be described in this way. Any set $\{e_i, e_{-i} \mid 1 \leq i \leq n\}$ of $2n$ isotropic vectors in V satisfying $f(e_i, e_j) = f(e_{-i}, e_{-j}) = 0$ and $f(e_i, e_{-j}) = \delta_{ij}$ $(1 \leq i, j \leq n)$ determines an apartment of Δ, namely

$$\Sigma(e_1, \ldots, e_n; e_{-1}, \ldots, e_{-n}) := \operatorname{Flag}\{\langle e_{i_1}, \ldots, e_{i_r} \rangle \mid i_j \neq \pm i_k$$
$$\text{for all } (1 \leq j \neq k \leq r)\}$$

Conversely, any apartment Σ of Δ is of this form (cf. [T1], §7).

Lemma 29 (cf. [Ab4], Lemma 1.2.3): $U, E \in X$ are opposite in Δ if and only if $V = U \oplus E^\perp$. $U \in \Delta^0(E)$ holds if and only if $U \pitchfork E^\perp$.

Proof: Choose an apartment $\Sigma = \Sigma(e_1, \ldots, e_n; e_{-1}, \ldots, e_{-n})$ containing U and E. We may assume $E = \langle e_1, \ldots, e_r \rangle$. Since op_Σ is type preserving, $\mathrm{op}_\Sigma(\langle e_i \rangle) = \langle e_j \rangle$ for some j. If $j \neq \pm i$, one easily finds a wall of Σ containing $\langle e_i \rangle$ but not $\langle e_j \rangle$. Therefore, $\mathrm{op}_\Sigma(\langle e_i \rangle) = \langle e_{-i} \rangle$ for all i, and hence $\mathrm{op}_\Sigma(E) = \langle e_{-1}, \ldots, e_{-r} \rangle =: E^0$. Observe that $E^0 \oplus E^\perp = V$ ($E^0 + E^\perp = V$ follows from $(E^0)^\perp \cap E = 0$ and Lemma 28) and that $W \cap E^\perp \neq 0$ or $W + E^\perp \neq V$ for all vertices $W \neq E^0$ of Σ. This implies the first claim and together with Lemma 16 i) of §1 also the second. $\qquad\square$

If $\mathcal{E}$ is a set of subspaces of V, we define as in §4
$X_\mathcal{E}(V) := \{U \in X \mid U \pitchfork \mathcal{E}\}$ and $T_\mathcal{E}(V) := \mathrm{Flag}\, X_\mathcal{E}(V)$. Note that $U \pitchfork E^\perp$ also yields $U \pitchfork E$ if $U, E \in X$ (in fact even $U \cap E = 0$ in view of Lemma 28). Hence Lemma 29 implies

Corollary 15: *For any simplex* $a = \{E_1 < \ldots < E_r\} \in \Delta$ *and*
$\mathcal{E}(a) := \{E_i, E_i^\perp \mid 1 \leq i \leq r\}$, $\Delta^0(a) = T_{\mathcal{E}(a)}(V)$. $\qquad\square$

We shall only consider subcomplexes of the form $T_\mathcal{E}(V)$ if $\mathcal{E}$ satisfies the following two conditions, where the first is obviously motivated by Lemma 28:

1. $\mathcal{E}$ is a finite subset of $\mathcal{W} = \mathcal{U} \cup \mathcal{U}^\perp$

2. $\mathcal{E} = \mathcal{E}^\perp$

Apart from being aesthetically appealing, the second condition is necessary in order to guarantee that the "upper links" in $T_\mathcal{E}(V)$ are again of the same form. To be more precise:

Given $U \in X$, we set $\overline{V} := U^\perp/U$ and denote by $\overline{f}$ (respectively $\overline{Q}$) the (σ, ε)-hermitian form (pseudo-quadratic form) induced by f (respectively Q) on $\overline{V} \times \overline{V}$ (respectively $\overline{V}$). Then the poset $X^{>U}$ is canonically isomorphic to $X(\overline{V})$. If additionally $U \in X_\mathcal{E}(V)$, we would like to obtain an induced isomorphism between $X_\mathcal{E}(V)^{>U}$ and $X_{\overline{\mathcal{E}}}(\overline{V})$ for $\overline{\mathcal{E}} := \{\overline{E} := (E \cap U^\perp) + U/U \mid E \in \mathcal{E}\}$. We therefore have to verify the following equivalence:

$$(*) \quad W \underset{V}{\pitchfork} E \iff W/U \underset{\overline{V}}{\pitchfork} \overline{E} \quad \text{for} \quad W \in X^{>U} \quad \text{and} \quad E \in \mathcal{E}$$

Now, **since** $\mathcal{E} = \mathcal{E}^\perp$, we do not only know $U \pitchfork E$ but also $U \pitchfork E^\perp$ and hence $U^\perp \pitchfork E$ by Lemma 28. Then $(*)$ is in fact easily checked. It is trivial for $U + E = V$ or

$U^\perp \cap E = 0$. If $U \cap E = 0$ and $U^\perp + E = V$, one obtains for $U < W < U^\perp$

$$W \cap E = 0 \iff W \cap E \subseteq U \iff W \cap (U^\perp \cap (E + U)) = U$$
$$\iff W/U \cap \overline{E} = 0 \,,$$
$$W + E = V \iff W + E \supseteq U^\perp \iff W + (U + (E \cap U^\perp)) = U^\perp$$
$$\iff W/U + \overline{E} = \overline{V}$$

If K is infinite, we shall need no other information about $\mathcal{E}$ than the two conditions stated above. However, if $K = \mathbb{F}_q$ is finite, we shall have to compare q with a certain natural number $N(\mathcal{E})$ associated to $\mathcal{E}$. Recall that non-degenerate anisotropic spaces over finite fields are at most 2-dimensional and therefore $m \leq 2n + 2$.

Definition 10:   *Set* $\mathcal{E}_j := \{E \in \mathcal{E} \mid \dim E = j\}$, $e_j := \# \mathcal{E}_j$ *and*

$$e_h^{(s)} := \sum_{j=0}^{2s} \binom{2s}{j} e_{h+j} \text{ for } h \in \mathbb{N}, \ s \in \mathbb{N}_0 \text{ and } h + 2s < m \,. \text{ We define}$$

$$N(\mathcal{E}) := e_1^{(n-1)} \qquad f \text{ is alternating } (\Rightarrow m = 2n) \text{ and } Q = 0 \,,$$
$$N(\mathcal{E}) := (e_1^{(n-1)})^2 \quad \text{if } m = 2n \text{ and } \sigma \neq \mathrm{id} \,,$$
$$N(\mathcal{E}) := 2\, e_2^{(n-1)} \quad \text{if } m = 2n + 1 \,, \text{ and finally}$$
$$N(\mathcal{E}) := \max\{e_2^{(n-1)} + e_3^{(n-1)} + 1,\ 2e_3^{(n-1)}\} \text{ if } m = 2n + 2$$

Below we shall need the following technical

Lemma 30:   *Given natural numbers* m, $1 \leq i < \frac{1}{2}(m - 2)$, $\overline{m} := m - 2i$ *and two finite sequences* $(e_1, \ldots, e_{m-1})$, $(\overline{e}_1, \ldots, \overline{e}_{\overline{m}-1})$ *contained in* $\mathbb{N}_0$. *Suppose that*

a) $\overline{e}_1 \leq e_1 + 2(e_2 + \ldots + e_{i+1}) + e_{i+2}$

b) $\overline{e}_j \leq e_{i+j-1} + 2e_{i+j} + e_{i+j+1}$ *for* $1 < j < \overline{m} - 1$

c) $\overline{e}_{\overline{m}-1} \leq e_{m-i-2} + 2(e_{m-i-1} + \ldots + e_{m-2}) + e_{m-1}$

Then $\overline{e}_h^{(s-i)} := \sum_{j=0}^{2(s-i)} \binom{2(s-i)}{j} \overline{e}_{h+j} \leq e_h^{(s)} := \sum_{j=0}^{2s} \binom{2s}{j} e_{h+j}$ *for all* $s, h \in \mathbb{N}$ *satisfying* $h + 2s < m$ *and* $i \leq s \leq \frac{1}{2}(m - 2)$.

Proof: The claim is verified by an elementary calculation using the identity

$$\binom{a}{b} + 2\binom{a}{b-1} + \binom{a}{b-2} = \binom{a+2}{b} \quad (a, b \in \mathbb{N}_0) \,. \qquad \square$$

85

Proposition 13: *Let $\mathcal{E}$ be a finite subset of $\mathcal{W}$ (cf. Lemma 28) satisfying $\mathcal{E} = \mathcal{E}^\perp$. Assume that either K is infinite or else $K = \mathbb{F}_q$ and $q \geq N(\mathcal{E})$. Then the subcomplex $T_\mathcal{E}(V)$ of the C_n building $\Delta = \mathrm{Flag}\, X$ is $(n-1)$-spherical.*

Proof: We want to apply the method explained in §3. Here the class **C** consists of all pairs described in §4, Proposition 12, all pairs $(T_{\mathcal{E}'}(V'), \mathrm{Flag}\, X')$ satisfying the conditions of the present section and joins of these pairs. We have to construct a filtration of $\kappa = T_\mathcal{E}(V)$ fulfilling the requirements of Lemma 22.

1. There exists a line $\ell \in X_\mathcal{E}(V) =: Y$

If K is infinite, this follows from §5, Lemma 25. If $K = \mathbb{F}_q$, the assertion is a consequence of Corollary 14, $q \geq N(\mathcal{E})$ and

$$N(\mathcal{E}) \geq e_1 + \ldots + e_{m-1} \qquad \text{if } \Lambda = \mathbb{F}_q \;(\Rightarrow f \text{ is alternating and } m = 2n),$$
$$N(\mathcal{E}) \geq (e_1 + \ldots + e_{m-1})^2 \quad \text{if } m = 2n \text{ and } \sigma \neq \mathrm{id},$$
$$N(\mathcal{E}) \geq 2(e_2 + \ldots + e_{m-1}) \quad \text{if } m = 2n + 1 \;,$$
$$N(\mathcal{E}) \geq e_2 + \ldots + e_{m-1} + 1 \;\; \text{if } m = 2n + 2 \;.$$

Note that an isotropic line $\ell \pitchfork \mathcal{E}_{m-1}$ automatically satisfies $\ell \pitchfork \mathcal{E}_{m-1}^\perp = \mathcal{E}_1$.

2. Description of $\kappa_0 = \mathrm{Flag}\, Y_0$

Recall that $U \in X$ is opposite to ℓ in Δ if and only if $U \cap \ell^\perp = 0$. If this is not the case, then the convex hull of U and ℓ in Δ contains exactly the (not necessarily distinct) vertices $U, U \cap \ell^\perp, (U \cap \ell^\perp) + \ell, \ell$ and consists of the edges generated by these vertices (consider an apartment $\Sigma = \Sigma(e_1, \ldots, e_n; e_{-1}, \ldots, e_{-n})$ containing U and ℓ). In view of the discussion in §3, our first candidate for Y_0 will therefore be the poset $Y_0' := \{ U \in Y \mid U \cap \ell^\perp, (U \cap \ell^\perp) + \ell \in Y \}$.

Now observe that $W + \ell \pitchfork E$ is equivalent to $W \pitchfork E + \ell$ for arbitrary subspaces W, E of V not containing ℓ since
$$(W + \ell) \cap E = 0 \iff W \cap E = 0 \text{ and } \ell \not\subseteq W + E \iff W \cap (E + \ell) = 0$$
Dually, for subspaces W, E of V not contained in $\ell^\perp$,
$$(W \cap \ell^\perp) + E = V \iff W + E = V \text{ and } (W \cap E) \not\subseteq \ell^\perp \iff W + (E \cap \ell^\perp) = V$$
which implies $W \cap \ell^\perp \pitchfork E \iff W \pitchfork E \cap \ell^\perp$.

Hence the conditions defining Y_0' may be rewritten in the following way: $U \in X$ is an element of Y_0' if and only if

i) $\ell \leq U$ and $U \pitchfork \mathcal{E}$ or

ii) $\ell \not\leq U \leq \ell^\perp$ and $U \pitchfork \mathcal{E} \cup (\mathcal{E} + \ell)$ or

iii)' $U \not\leq \ell^\perp$, $\dim U > 1$ and $U \pitchfork \mathcal{E} \cup (\mathcal{E} \cap \ell^\perp) \cup ((\mathcal{E} \cap \ell^\perp) + \ell)$

In view of the relative links we have to consider below, we replace iii)' by the following condition

iii) $U \not\leq \ell^\perp$, $\dim U > 1$ and $U \pitchfork \mathcal{E} \cup (\mathcal{E} \cap \ell^\perp) \cup (\mathcal{E} + \ell) \cup ((\mathcal{E} \cap \ell^\perp) + \ell)$

and define $Y_0 := \{U \in X \mid U \text{ satisfies i), ii) or iii)}\}$.

Observe that $|\kappa_0| = |\operatorname{Flag} Y_0|$ can be contracted onto ℓ via
$U \longmapsto U \cap \ell^\perp \longmapsto (U \cap \ell^\perp) + \ell \longmapsto \ell$. The contractibility of $|\kappa_0|$ can also be proved without using convex hulls. Start with $\operatorname{st}_\kappa(\ell)$. Add successively in a dimension-increasing order the $U \in X$ fulfilling ii). Since the occurring relative links are contractible (onto $U + \ell$, respectively),
$|\operatorname{Flag}\{U \in X \mid U \text{ satisfies i) or ii)}\}|$ is also contractible. Apply then a dimension-decreasing filtration in order to deduce that $|\operatorname{Flag} Y_0|$ is contractible.

3. A dimension-increasing filtration

The first part of the filtration of κ is defined as follows: Set

$Z \;:=\; \{U \in Y \mid \ell \leq U \text{ or } U \pitchfork \mathcal{E} + \ell\}$,

$Y_i \;:=\; \{U \in Z \mid U \in Y_0 \text{ or } \dim U \leq i\}$ and $\kappa_i := \operatorname{Flag} Y_i \; (1 \leq i \leq n)$

For arbitrary given $U \in Y_i \setminus Y_{i-1}$, we have to study $\kappa_{i-1}^{<U} = \operatorname{Flag} Z^{<U}$ and $\kappa_{i-1}^{>U} = \operatorname{Flag} Y_0^{>U}$. Note that $\dim U = i$, $\ell \not\leq U \not\leq \ell^\perp$ and $U \pitchfork \mathcal{E} + \ell$.

a) Obviously, $\Delta^{<U} = \operatorname{Flag}\{0 < W < U\}$ is the A_{i-1} building associated to U . Since $U \underset{V}{\pitchfork} \mathcal{E} \cup (\mathcal{E} + \ell)$, $W \underset{V}{\pitchfork} F$ if and only if $W \underset{U}{\pitchfork} (F \cap U)$ for any $W < U$ and $F \in \mathcal{E} \cup (\mathcal{E} + \ell)$ (cf. step 3 in the proof of Proposition 12). Setting $\mathcal{E}' := \{F \cap U \mid F \in \mathcal{E} \cup (\mathcal{E} + \ell)\}$, one thus obtains $Z^{<U} = \{0 < W < U \mid W \underset{U}{\pitchfork} \mathcal{E}'\}$.
In case $K = \mathbb{F}_q$, we note that $e'_j := \#\mathcal{E}'_j \leq e_{j+m-i} + e_{j-1+m-i}$ which implies

$$\sum_{j=1}^{i-1} \binom{i-2}{j-1} e'_j \leq \sum_{j=m-i}^{m-1} \binom{i-1}{j-(m-i)} e_j \leq N(\mathcal{E}) \leq q \ .$$

Hence the pair $(\kappa_{i-1}^{<U}, \Delta^{<U})$ is of the form described in Proposition 12.

b) As already mentioned while motivating the condition $\mathcal{E} = \mathcal{E}^\perp$, $\Delta^{>U}$ is canonically isomorphic to the C_{n-i} building Flag $X(\overline{V})$ associated to the quadratic space $\overline{V} := U^\perp/U$. We assume $i < n$ here since we need not study the empty set.

Set $\quad \mathcal{F} \; := \; \mathcal{E} \cup (\mathcal{E} \cap \ell^\perp) \cup (\mathcal{E} + \ell) \cup ((\mathcal{E} \cap \ell^\perp) + \ell)$

and $\quad \overline{\mathcal{E}} \; := \; \{\overline{F} := (F \cap U^\perp) + U/U \mid F \in \mathcal{F}\}$.

We want to show $Y_0^{>U} \cong X_{\overline{\mathcal{E}}}(\overline{V})$. In the following, we always assume $F \in \mathcal{F}$ and $E \in \mathcal{E}$.

(1) There exist $F', F'' \in \mathcal{F}$ satisfying $F' \le F \le F''$, $\operatorname{codim}_F F' \le 1$, $\operatorname{codim}_{F''} F \le 1$ as well as $U \pitchfork F''$ and $U^\perp \pitchfork F'$.

   If $F = E$ or $E \cap \ell^\perp$, choose $F'' = E$.
   If $F = E + \ell$ or $(E \cap \ell^\perp) + \ell$, choose $F'' = E + \ell$.
   If $F = E$ or $E + \ell$, choose $F' = E$.
   If $F = E \cap \ell^\perp$ or $(E \cap \ell^\perp) + \ell$, choose $F' = E \cap \ell^\perp$.

(2) $\dim F > i$ implies $U^\perp + F = V$

   Choose F' as in (1) and observe $U^\perp + F' = V$.

(3) $\operatorname{codim}_V F > i$ implies $U \cap F = 0$

   Choose F'' as in (1) and observe $U \cap F'' = 0$.

(4) $\dim F \le i$ implies either   i) $U^\perp \cap F = 0$ or else
   ii) $\dim(U^\perp \cap F) = 1$, $U \cap F = 0$ and

   $$F \in ((\mathcal{E}_1 \cup \ldots \cup \mathcal{E}_{i-1}) + \ell) \cup (((\mathcal{E}_2 \cup \ldots \cup \mathcal{E}_i) \cap \ell^\perp) + \ell)$$

   If $F = E$ or $E \cap \ell^\perp$, i) follows from $U^\perp \pitchfork F$. If $F = \ell$, $U^\perp \cap F = 0$ since $U \not\subseteq \ell^\perp$. If $F = E + \ell$ or $F = (E \cap \ell^\perp) + \ell \ne \ell$, we see by choosing F' as in (1) that $\dim(U^\perp \cap F) \le 1$. Finally, $U \cap F = 0$ follows from $i < n < \operatorname{codim}_V F$ and (3).

(5) $\operatorname{codim}_V F \le i$ implies either i) $U + F = V$ or else
   ii) $\operatorname{codim}_V(U + F) = 1$, $U^\perp + F = V$ and
   $F = E \cap \ell^\perp$, $\operatorname{codim}_V E \le i - 1$ or $F = (E \cap \ell^\perp) + \ell$, $1 < \operatorname{codim}_V E \le i$.
   The proof is dual to that of statement (4).

(6) For any subspace W satisfying $U < W < U^\perp$, $W \underset{V}{\pitchfork} F$ is equivalent to $W/U \underset{\overline{V}}{\pitchfork} \overline{F}$.

This is obvious if $U^\perp \cap F = 0$ or $U + F = V$. In the situation described in (4) ii), one obtains

$$W \underset{V}{\pitchfork} F \iff U^\perp \cap F \nleq W \iff W \cap (U^\perp \cap F) = 0$$
$$\iff W \cap ((U^\perp \cap F) + U) = U$$
$$\iff W/U \cap \overline{F} = 0 \iff W/U \underset{\overline{V}}{\pitchfork} \overline{F}$$

Dually, in the situation described in (5) ii), it holds

$$W \underset{V}{\pitchfork} F \iff W \nleq U + F \iff W/U + \overline{F} = \overline{V} \iff W/U \underset{\overline{V}}{\pitchfork} \overline{F}$$

Assume $\dim F > i$ and $\operatorname{codim}_V F > i$ at last. Then $U^\perp + F = V$ and $U \cap F = 0$ by (2) and (3). But in this case, we already observed that $W \underset{V}{\pitchfork} F$ and $W/U \underset{\overline{V}}{\pitchfork} \overline{F}$ are equivalent (cf. the proof of $(*)$ below Corollary 15).

Since $Y_0^{>U} = \{ U < W < U^\perp \mid W \in X \text{ and } W \pitchfork \mathcal{F} \}$, (6) implies

(7) $Y_0^{>U}$ is canonically isomorphic to the poset $X_{\overline{\mathcal{E}}}(\overline{V})$

(8) $\overline{\mathcal{E}}^\perp = \overline{\mathcal{E}}$ and $\dim \overline{F} < \infty$ or $\dim \overline{F}^\perp < \infty$ for any $\overline{F} \in \overline{\mathcal{E}}$.
This follows from Lemma 28 and $\mathcal{F}^\perp = \mathcal{F} \subseteq \mathcal{W}$.

If $K = \mathbb{F}_q$, we also have to estimate $N(\overline{\mathcal{E}})$. (2) and (3) imply

(9) If $\dim F > i$ and $\operatorname{codim}_V F > i$, then $\dim \overline{F} = \dim F - i$.

Now (4), (5) and (9) yield the following statement:

(10)   i) Assume that $\overline{m} := m - 2i$ is greater than 2. Then it follows

$$\begin{aligned}
\overline{e}_1 &:= \#\overline{\mathcal{E}}_1 \leq e_1 + 2(e_2 + \ldots + e_{i+1}) + e_{i+2} , \\
\overline{e}_j &:= \#\overline{\mathcal{E}}_j \leq e_{i+j-1} + 2e_{i+j} + e_{i+j+1} \quad \text{for } 1 < j < \overline{m} - 1, \\
\overline{e}_{\overline{m}-1} &:= \#\overline{\mathcal{E}}_{\overline{m}-1} \leq e_{m-i-2} + 2(e_{m-i-1} + \ldots + e_{m-2}) + e_{m-1}
\end{aligned}$$

   ii) If $\overline{m} = 2$ ($\iff m = 2n$ and $i = n - 1$) , then
   $$\overline{e}_1 \leq e_1 + 2(e_2 + \ldots + e_{m-2}) + e_{m-1}$$

Applying Lemma 30 for $s = n - 1$ and Definition 10 (note that $n - i$ is the Witt index of $\overline{V}$), one derives from (10)

89

(11) $N(\overline{\mathcal{E}}) \leq N(\mathcal{E}) \leq q$ if $K = \mathbb{F}_q$

Statements (7),(8) and (11) show that

$$(\kappa_{i-1}^{>U}, \Delta^{>U}) = (\text{Flag } Y_0^{>U}, \text{Flag } X^{>U}) \cong (T_{\overline{\mathcal{E}}}(\overline{V}), \text{Flag } X(\overline{V})) \in \mathbf{C}$$

4. A dimension-decreasing filtration

The second part of the filtration of $\kappa = \text{Flag } Y$ is defined by

$Y_i := \{U \in Y \mid U \in Z \text{ or } \dim U \geq 2n + 1 - i\}$ and

$\kappa_i := \text{Flag } Y_i$ for $n \leq i \leq 2n$.

Obviously, $Y_n = Z$ as above and $Y_{2n} = Y$. We fix an arbitrary

$U \in Y_i \setminus Y_{i-1}$ $(n + 1 \leq i \leq 2n)$ and set $k := 2n + 1 - i = \dim U$. Note that

$\ell \not\leq U$, $Y_{i-1}^{<U} = Z^{<U} = \{0 < W < U \mid W \underset{V}{\pitchfork} \mathcal{E} \cup (\mathcal{E} + \ell)\}$ and $Y_{i-1}^{>U} = Y^{>U}$.

a) Everything we need to know of $Z^{<U}$ was already derived in Step 3 of the proof

of Proposition 12. Firstly, since $U \underset{V}{\pitchfork} \mathcal{E}$, $W \underset{V}{\pitchfork} F$ is equivalent to $W \underset{U}{\pitchfork} (F \cap U)$ for

any $W < U$ and $F \in \mathcal{E} \cup (\mathcal{E} + \ell)$. Setting $\mathcal{E}' := (\mathcal{E} \cap U) \cup ((\mathcal{E} + \ell) \cap U)$, we thus

obtain $Z^{<U} = \{0 < W < U \mid W \underset{U}{\pitchfork} \mathcal{E}'\}$. Secondly, if $K = \mathbb{F}_q$ and in particular

$\dim V = m < \infty$, we know that $\sum_{j=1}^{k-1} \binom{k-2}{j-1} e_j' \leq e_1 + \ldots + e_{m-k-1} + \sum_{r=m-k}^{m-1} \binom{k-1}{r-(m-k)} e_r$.

Now the term on the right side of this inequality is $\leq N(\mathcal{E})$. This follows from

$e_1 = e_{m-1}$, $e_2 = e_{m-2}$, $n \geq 2$ and $\binom{k-1}{r-(m-k)} \leq \binom{m-a-1}{r-a}$ for any integer $0 \leq a \leq m - k$.

Note that this is the only place where we make use of $N(\mathcal{E}) \geq 2e_3^{(n-1)}$ for $m = 2n + 2$.

It is proved now that $(\kappa_{i-1}^{<U}, \Delta^{<U}) = (\text{Flag } Z^{<U}, \text{Flag}\{0 < W < U\}) \in \mathbf{C}$.

b) Since $\mathcal{E} = \mathcal{E}^{\perp}$ and $U \underset{V}{\pitchfork} \mathcal{E}$, $Y^{>U}$ is canonically isomorphic to $X_{\overline{\mathcal{E}}}(\overline{V})$ for $\overline{V} := U^{\perp}/U$

and $\overline{\mathcal{E}} := \{\overline{E} := (E \cap U^{\perp}) + U/U \mid E \in \mathcal{E}\}$. In this situation, $N(\overline{\mathcal{E}}) \leq N(\mathcal{E})$ is obvious

for finite K . Hence $(\kappa_{i-1}^{>U}, \Delta^{>U}) \cong (T_{\overline{\mathcal{E}}}(\overline{V}), \text{Flag } X(\overline{V})) \in \mathbf{C}$. □

Corollary 16: *Let $\Delta = \text{Flag } X(V)$ be a classical C_n building as described above.*
Assume that either K is infinite or else $K = \mathbb{F}_q$ and

$$q \geq 2^{2n-2} \text{ if } f \text{ is alternating and } Q = 0 ,$$

$$q \geq 2^{4n-4} \text{ if } m = 2n \text{ and } \sigma \neq \text{id} ,$$

$$q \geq 2^{2n-1} \text{ if } m = 2n + 1 \text{ or } 2n + 2 .$$

Then $\Delta^0(a)$ is $(n - 1)$-spherical for any $a \in \Delta$. □

Remark 14: Though it is more difficult to deal with classical C_n buildings than with A_n buildings, the result obtained above is exactly of the same form as in the A_n case: Δ possesses the property (S_Δ) if every panel is contained in at least $2^{\varphi(n)} + 1$ chambers, where φ is a linear function of $n = \mathrm{rk}\,\Delta$. It has been tried without success to improve this result for A_n buildings. Therefore, I did not dwell on looking for better bounds while proving Proposition 13. Even in Corollary 16, the requirement $q \geq 2^{2n-1}$ can be resplaced by $q \geq 2(2^{2n-2} - \binom{2n-2}{n-2})$ for $m = 2n+2$ since $e_{n+1} = 0$ for $\mathcal{E} = \mathcal{E}(a)$ as in Corollary 15. Furthermore, one can study directly the connectedness of $T_\mathcal{E}(V)$ in the C_2 case and then start the induction with $n = 2$ instead of $n = 1$. For C_3 buildings, this leads to part ii) of Proposition 11 in §2.

§ 7   The case D_n

In this section, we specialize the notations and agreements of §5 in the following way: K is a (commutative) field, $\sigma = \mathrm{id}_K$, $\varepsilon = 1$, $\Lambda = 0$, V a K-vector space of dimension $m = 2n \geq 4$, $Q : V \longrightarrow K$ an ordinary quadratic form and $f : V \times V \longrightarrow K$ the symmetric bilinear form associated to Q. We assume that V is non-degenerate and of Witt index n. Hence (V, Q, f) is the **hyperbolic space of dimension $2n$ over** K which is unique up to isomorphism. Set again

$X := X(V) := \{0 < U < V \mid U \text{ is a totally isotropic subspace of } V\}$.

Recall that $\Delta := \mathrm{Flag}\,X$ is a weak C_n building. One obtains a thick D_n building by defining $\quad \widetilde{X} := \widetilde{X}(V) := \{U \in X \mid \dim U \neq n - 1\}$ and $\widetilde{\Delta} := \mathrm{Orifl}\,\widetilde{X}$,

"Orifl" denoting the "oriflamme complex" in the sense of [T1], §7.12. This means that $\widetilde{\Delta}$ is the flag complex associated to the following incidence relation on $\widetilde{X}$:

$$U I W :\Leftrightarrow U \subseteq W \text{ or } U \supseteq W \text{ or } \dim(U \cap W) = n - 1 \ (U, W \in \widetilde{X})$$

For $n \geq 4$, every thick D_n building is of the form $\widetilde{\Delta}$ for some field K according to [T1], Proposition 8.4.3. We also admit $n = 2$ and $n = 3$ here which leads to certain $D_2 = A_1 \times A_1$ and $D_3 = A_3$ buildings. As in §6, any hyperbolic basis $\{e_i, e_{-i} \mid 1 \leq i \leq n\}$ of V determines an apartment $\Sigma(e_1, \ldots, e_n; e_{-1}, \ldots, e_{-n})$ of Δ. Furthermore,

$$\widetilde{\Sigma}(e_1, \ldots, e_n; e_{-1}, \ldots, e_{-n}) := \mathrm{Orifl}\,\{\langle e_{i_1}, \ldots, e_{i_r}\rangle \mid 1 \leq r \leq n, \ r \neq n - 1;$$
$$i_j \neq \pm i_k \text{ for all } 1 \leq j \neq k \leq r\}$$

is an apartment of $\widetilde{\Delta}$, and every apartment $\widetilde{\Sigma}$ of $\widetilde{\Delta}$ is of this form (cf. [T1], §7.12).

In the following, we need a modification of the notion of transversality.

Definition 11: *For arbitrary subspaces U, W of V , we define*

$$U \tilde{\pitchfork} W :\Leftrightarrow U \underset{V}{\pitchfork} W \text{ or } (U, W \in \widetilde{X}, \ U = U^{\perp}, \ W = W^{\perp} \text{ and } \dim(U \cap W) = 1)$$

For any set $\mathcal{E}$ of subspaces of V , one sets (deviating from Definition 9)

$$X_{\mathcal{E}}(V) := \{U \in X \mid U \tilde{\pitchfork} \mathcal{E}\}, \quad T_{\mathcal{E}}(V) := \text{Flag } X_{\mathcal{E}}(V)$$
$$\widetilde{X}_{\mathcal{E}}(V) := \{U \in \widetilde{X} \mid U \tilde{\pitchfork} \mathcal{E}\}, \quad \widetilde{T}_{\mathcal{E}}(V) := \text{Orifl } \widetilde{X}_{\mathcal{E}}(V)$$

Lemma 31 (cf. [Ab4], 1.2.6 and 1.2.8):

i) $U, E \in \widetilde{X}$ *are opposite in $\widetilde{\Delta}$ if and only if $V = U \oplus E^{\perp}$.*

ii) *For $U, E \in \widetilde{X}$, $U \in \widetilde{\Delta}^{0}(E)$ is equivalent to $U \tilde{\pitchfork} E^{\perp}$.*

iii) *Assume $U, E \in X$, $\dim U = n - 1$ and $U = U_1 \cap U_2$ for $U_1, U_2 \in \widetilde{X}$. Then $U \pitchfork E^{\perp}$ is equivalent to $U_1 \tilde{\pitchfork} E^{\perp}$ and $U_2 \tilde{\pitchfork} E^{\perp}$.*

Proof: Choose a hyperbolic basis of V such that
$U, E \in \widetilde{\Sigma} = \widetilde{\Sigma}(e_1, \dots, e_n; e_{-1}, \dots, e_{-n})$ in i) and ii) and
$U, E \in \Sigma = \Sigma(e_1, \dots, e_n; e_{-1}, \dots, e_{-n})$ in iii).

i) Assume $E = \langle e_1, \dots, e_r \rangle$. As in Lemma 29, $\dim \text{op}_{\widetilde{\Sigma}}(\langle e_i \rangle) = 1$ implies $\text{op}_{\widetilde{\Sigma}}(\langle e_i \rangle) = \langle e_{-i} \rangle$, $\text{op}_{\widetilde{\Sigma}}(E) = \langle e_{-1}, \dots, e_{-r} \rangle =: E^0$ and hence the claim.

ii) By Lemma 16 i), $U \in \widetilde{\Delta}^0(E)$ if and only if U and $\text{op}_{\widetilde{\Sigma}}(E) = E^0$ are incident in $\widetilde{X}$. This implies ii).

iii) Note that $U \in \Sigma$ implies $U_1 \in \Sigma$ and $U_2 \in \Sigma$. For $\dim E < n$, the claim easily follows from $E \in \Sigma$. For $E = E^{\perp}$, one uses
$\dim(U_1 \cap E) \not\equiv \dim(U_2 \cap E) \bmod 2$ in order to deduce $U_1 \cap E = 0$ or $U_2 \cap E = 0$
from $U_1 \tilde{\pitchfork} E$ and $U_2 \tilde{\pitchfork} E$. $\qquad \square$

Corollary 17: *Let* $a = \{E_1, \ldots, E_r\}$ *be a simplex of* $\tilde{\Delta} = \mathrm{Orifl}\,\widetilde{X}$ *and set* $\mathcal{E}(a) := \{E_i, E_i^\perp \mid 1 \leq i \leq r\}$. *Then it follows*

i) $\tilde{\Delta}^0(a) = \tilde{T}_{\mathcal{E}(a)}(V)$.

ii) $|\tilde{\Delta}^0(a)|$ *is homeomorphic to* $|T_{\mathcal{E}(a)}(V)|$.

Proof: Note that $U \pitchfork E_i^\perp$ implies $U \pitchfork E_i$ if $U \in X$.

i) follows from Lemma 16 and Lemma 31.

ii) Recall that Δ is a simplicial subdivision of $\tilde{\Delta}$ obtained by cutting in two the chamber of $\tilde{\Delta}$. One introduces one new vertex $U_1 \cap U_2$ on every edge $\{U_1, U_2\}$ of $\tilde{\Delta}$ with $U_1 = U_1^\perp$ and $U_2 = U_2^\perp$. This vertex is joined to any simplex of $\mathrm{st}_{\tilde{\Delta}}\{U_1, U_2\}$ not containing $\{U_1, U_2\}$. The resulting simplicial complex is easily identified with $\mathrm{Flag}\,X = \Delta$.

Now by Lemma 31 iii), the simplicial subdivision obtained from $\tilde{T}_{\mathcal{E}(a)}(V)$ is precisely $\mathrm{Flag}\,X_{\mathcal{E}(a)}(V)$, implying of course ii). $\qquad\square$

Remark 15: I want to stress that $|\tilde{\Delta}^0(a)|$ is **not** homeomorphic to $|\Delta^0(a)|$ in case $a = \{E_1, \ldots, E_r\} \in \Delta \cap \tilde{\Delta}$ contains a vertex, say E_r , satisfying $E_r = E_r^\perp$. In fact, Lemma 29 (which also holds for weak C_n buildings) shows that $\Delta^0(a) = \mathrm{Flag}\{U \in X \mid U \pitchfork \mathcal{E}(a)\}$. In particular, given $U \in \Delta^0(a)$ with $\dim U = n-1$, only one of the two maximal totally isotropic subspaces U_1, U_2 containing U can be an element of $\Delta^0(a)$ in view of $\dim(U_1 \cap E_r) \not\equiv \dim(U_2 \cap E_r) \bmod 2$. It follows that $|\Delta^0(a)|$ is homotopy equivalent to the $(n-2)$-dimensional space $|\mathrm{Flag}\{U \in \widetilde{X} \mid U \pitchfork \mathcal{E}(a)\}|$ which can be shown to be non-contractible. On the other side, we are going to prove now that $\tilde{\Delta}^0(a)$ is $(n-1)$-spherical provided that $\#K \geq 2^{2n-1}$.

Before proceeding, we have to decide whether we shall work with thick D_n buildings or with weak C_n buildings in the following. Both approaches lead to technical difficulties which cannot be discussed in detail here. I just remind the reader of Problem iv) mentioned in §3. Suppose we are given a full subcomplex κ of $\tilde{\Delta}$, a line (= vertex of type 1) $\ell \in \kappa$ and an edge $\{U_1, U_2\} \in \kappa$ satisfying $U_1 = U_1^\perp$ and $U_2 = U_2^\perp$. If the convex hulls of ℓ and U_1 and of ℓ and U_2 are included in κ , we still do not know whether the convex hull of ℓ and $\{U_1, U_2\}$ is contained in κ. This problem leads to considering the center $U_1 \cap U_2$ of $\{U_1, U_2\}$ as well, thus indicating that one should study weak C_n buildings instead of thick D_n buildings here. However, the definition of $\tilde{\pitchfork}$ comes from the D_n structure and produces new difficulties now. The most important is the following:

Given $U \in X_{\mathcal{E}}(V)$, we again want to identify $X_{\mathcal{E}}(V)^{>U}$ with $X_{\overline{\mathcal{E}}}(\overline{V})$, where $\overline{V} := U^{\perp}/U$, $\overline{\mathcal{E}} := \{\overline{E} := (E \cap U^{\perp}) + U/U \mid E \in \mathcal{E}\}$. But it may happen that $E \in \mathcal{E}$ is not maximal totally isotropic whereas $\overline{E}$ is maximal totally isotropic in $\overline{V}$. In this case, $W \tilde{\underset{V}{\pitchfork}} E$ is not equivalent to $W/U \tilde{\underset{\overline{V}}{\pitchfork}} \overline{E}$ for $W \in X^{>U}$. Therefore, we have to introduce further restrictions which, by the way, cannot be translated into transversality conditions in general.

Definition 12: *Let U and E be subspaces of V, $\dim U < n$ and $\dim E = n$. Assume that E is not totally isotropic. Then we define*

$$U @ E :\Leftrightarrow U^{\perp} \cap E \text{ is not totally isotropic}$$

We list some facts concerning the relation $@$. Set

$$\mathcal{U}_n := \mathcal{U}_n(V) := \{A < V \mid \dim A = n \text{ and } A \notin X\} \text{ and}$$
$$\mathcal{M}(A) := \{M < A \mid \dim M = n - 1 \text{ and } M \in X\} \text{ for any } A \in \mathcal{U}_n.$$

Lemma 32: *Let subspaces U, E of V be given. Assume $U \in X$, $\dim U < n$ and $U \pitchfork E, E^{\perp}$. Denote by $\overline{V}$ the hyperbolic K-space $\overline{V} := U^{\perp}/U$ of Witt index $\overline{n} := n - \dim U$, and set $\overline{E} := (U^{\perp} \cap E) + U/U$. Then it holds:*

i) $E \in \mathcal{U}_n$ *and* $U @ E$ *imply* $\overline{E} \in \mathcal{U}_{\overline{n}} = \mathcal{U}_{\overline{n}}(\overline{V})$

ii) *Provided that* $U @ E$ *in case* $E \in \mathcal{U}_n$,

$$W \tilde{\underset{V}{\pitchfork}} E \quad \text{es equivalent to} \quad W/U \tilde{\underset{\overline{V}}{\pitchfork}} \overline{E} \quad \text{for any} \quad W \in X^{>U}.$$

Note that the latter is always true by definition if $\dim U = n - 1$.

In the following statements, we suppose that $E \in \mathcal{U}_n$.

94

iii) *For $W < U$, $U @ E$ implies $W @ E$. For $W \in X^{>U}$ and $U @ E$,*

$W \underset{V}{@} E$ *if and only if* $W/U \underset{\overline{V}}{@} \overline{E}$

iv) *$U @ E$ is equivalent to $U @ E^{\perp}$*

v) *If $\dim U = 1$, $U @ E$ is equivalent to $U \not\leq M^{\perp}$ for all $M \in \mathcal{M}(E)$*

vi) *$\# \mathcal{M}(E) \leq 2$*

Proof:

i) Since $U \cap E = 0$ and $U^{\perp} + E = V$, $\dim \overline{E} = \dim E - \dim U = \overline{n}$.
   Since $U @ E$, $U^{\perp} \cap E$ and $\overline{E}$ are not totally isotropic.

ii) Below Corollary 15 in §6, we already showed $W \underset{V}{\pitchfork} E \iff W/U \underset{\overline{V}}{\pitchfork} \overline{E}$. This
   proves our claim for $\dim E \neq n$ and in view of i) also for $E \in \mathcal{U}_n$. If E is
   maximal totally isotropic, the same is true for $\overline{E}$, and the assertion follows
   from $\dim(W \cap E) = \dim(W/U \cap \overline{E})$.

iii) The first part is trivial and the second follows from
   $(W/U)^{\perp} \cap \overline{E} = (W^{\perp} \cap E) + U/U$.

iv) Observe again that $\dim U + (U^{\perp} \cap E) = n = \dim U + (U^{\perp} \cap E^{\perp})$ because of
   $U \pitchfork E, E^{\perp}$. If $U^{\perp} \cap E$ is totally isotropic, this implies in particular
   $U^{\perp} \cap E \subseteq U + E^{\perp}$ and hence $U + (U^{\perp} \cap E) = U + (U^{\perp} \cap E^{\perp})$. Therefore,
   $U^{\perp} \cap E^{\perp}$ is totally isotropic as well.

v) Since $\dim(U^{\perp} \cap E) = n - 1$, $U @ E$ is equivalent to $U^{\perp} \cap E \neq M$ for all
   $M \in \mathcal{M}(E)$. Now observe
   $U^{\perp} \cap E \neq M \iff U + E^{\perp} \neq M^{\perp} \iff U \not\leq M^{\perp}$.

vi) Set $E_0 := E \cap E^{\perp} \cap Q^{-1}(0)$ and $n_0 := \dim E_0$. If $n_0 < n - 2$, the Witt index of
   E is $< n - 1$ and $\mathcal{M}(E) = \emptyset$. If $n_0 = n - 2$ and $E = E_0 \oplus E_1$, $\# \mathcal{M}(E) = 0$
   or 2 according as E_1 is anisotropic or a hyperbolic plane. Finally, if $n_0 = n - 1$,
   then $\mathcal{M}(E) = \{E_0\}$. $\square$

Before going to analyse subcomplexes of type $T_{\mathcal{E}}(V)$, we need two technical lemmata concerning certain relative links occurring in the proof of Proposition 14 below.

The reader is supposed to be familiar with the general method of the sphericity proofs as well as with the details of the preceding sections. Therefore, I will concentrate on the essential arguments now.

Lemma 33: *Assume $U \in X$ and $k := \dim U \geq 2$. Let $\mathcal{E}$ be a finite set of subspaces of U and set $e_j := \# \mathcal{E}_j$ $(1 \leq j \leq k-1)$. Let further $\mathcal{F} = \mathcal{F}_1 \supseteq \mathcal{F}_2 \supseteq \ldots \supseteq \mathcal{F}_{k-1}$ be finite subsets of $\mathcal{U}_n(V)$ satisfying*

(1) $U \cap F \in \mathcal{E} \cup \{0\}$ *for all* $F \in \mathcal{F}$

(2) $\mathcal{F}_j^\perp = \mathcal{F}_j$

(3) $\dim(U \cap F) \leq k - 1 - j$ *for all* $F \in \mathcal{F}_j$ $(1 \leq j \leq k-1)$

Set $X_{\mathcal{E};\mathcal{F}}(U;V) := \{0 < W < U \mid W \underset{U}{\pitchfork} \mathcal{E}$ *and* $W \underset{V}{@} \mathcal{F}_j$ *for* $\dim W = j\}$.

Finally, assume $\#K \geq \sum\limits_{j=1}^{k-1} \binom{k-2}{j-1} e_j + 2^{k-1} s$, *where* $s := \# \mathcal{F}$. *Then* Flag $X_{\mathcal{E};\mathcal{F}}(U;V)$ *is* $(k-2)$-*spherical.*

Proof: 1. There exists a line $\ell \in X_{\mathcal{E};\mathcal{F}}(U;V) =: Y$.

For any $f \in \mathcal{F} = \mathcal{F}_1 = \mathcal{F}_1^\perp$ and $M \in \mathcal{M}(F)$, U is not contained in $M^\perp$ since otherwise $\dim(U \cap F^\perp) \geq k-1$. Therefore,
$(\mathcal{E} \setminus \{0, U\}) \cup \{M^\perp \cap U \mid, M \in \mathcal{M}(F), F \in \mathcal{F}\}$ is a finite set of proper subspaces of U. Its cardinality is $\leq e_1 + \ldots + e_{k-1} + 2s \leq \#K$. Hence there exists a line $\ell < U$ not contained in any of these subspaces.
By Lemma 32 (note also $\ell \pitchfork \mathcal{E} \Rightarrow \ell \pitchfork (\mathcal{F} \cap U) \Rightarrow \ell \pitchfork \mathcal{F})$, $\ell @ \mathcal{F}$.

2. Description of Y_0

Assume $A \in Y$, $\ell \not\leq A$ and $j = \dim A < k-1$. Recall that $A + \ell \underset{U}{\pitchfork} \mathcal{E}$ is equivalent to $A \underset{U}{\pitchfork} \mathcal{E} + \ell$ (cf. §4). For any $F \in \mathcal{F}_{j+1}$, we obtain

$$A + \ell \; @ \; F \quad \Longleftrightarrow \quad A^\perp \cap \ell^\perp \cap F \notin X \quad \Longleftrightarrow \quad \ell + (A^\perp \cap F \cap \ell^\perp) \notin X$$
$$\Longleftrightarrow \quad A^\perp \cap ((F \cap \ell^\perp) + \ell) \notin X \quad \Longleftrightarrow \quad A \; @ \; (F \cap \ell^\perp) + \ell =: F'$$

Note that $F' \in \mathcal{U}_n(V)$ since $\ell \underset{V}{\pitchfork} F, F^\perp$ and $\ell @ F$ and that $(F')^\perp = (F^\perp)'$. We have proved now

$$A + \ell \in Y \quad \Longleftrightarrow \quad A \underset{U}{\pitchfork} \mathcal{E} + \ell \quad \text{and} \quad A \underset{V}{@} (\mathcal{F}_{j+1} \cap \ell^\perp) + \ell$$

Set $\mathcal{F}'_j := \mathcal{F}_j \cup ((\mathcal{F}_{j+1} \cap \ell^\perp) + \ell)$ for $1 \le j \le k-2$, $\mathcal{F}' := \mathcal{F}'_1$ and

$$
\begin{aligned}
Y_0 &:= \{A \in Y \mid \ell \le A \text{ or } (A \underset{U}{\pitchfork} \mathcal{E} + \ell, \ j = \dim A < k-1 \text{ and } A \underset{V}{@} \mathcal{F}'_j)\} \\
&= \{A \in Y \mid A + \ell \in Y\}
\end{aligned}
$$

Then $|\operatorname{Flag} Y_0|$ can be contracted onto ℓ .

3. The poset $Y_0^{<W}$

Here we assume $W \in Y$, $\ell \not\le W$, $2 \le \dim W = k - i \le k-1$ and additionally $W \underset{U}{\pitchfork} (\mathcal{F} \cap U) + \ell$. We set $\mathcal{E}' := (\mathcal{E} \cap W) \cup ((\mathcal{E} + \ell) \cap W)$. As in §4, $W \underset{U}{\pitchfork} \mathcal{E}$ implies

$$
A \underset{U}{\pitchfork} \mathcal{E} \cup (\mathcal{E} + \ell) \iff A \underset{W}{\pitchfork} \mathcal{E}' \text{ for any } A < W . \text{ Recall also that}
$$

$$
\sum_{j=1}^{k-i-1} \binom{k-i-2}{j-1} e'_j \le \sum_{j=1}^{k-1} \binom{k-2}{j-1} e_j \quad \text{for} \quad e'_j := \# \mathcal{E}'_j
$$

Next we observe that all the conditions stated in the lemma are satisfied with $(U; \mathcal{E}; \mathcal{F}_j, \ 1 \le j \le k-1)$ replaced by $(W; \mathcal{E}'; \mathcal{F}'_j, \ 1 \le j \le k - i - 1)$.

(1) follows from $W \cap F = W \cap (U \cap F)$ and
$W \cap ((F \cap \ell^\perp) + \ell) = W \cap (F + \ell) = W \cap ((U \cap F) + \ell)$. (2) is obvious.

As for (3) , we first note that $W \underset{U}{\pitchfork} (\mathcal{F}_j \cap U)$ implies

$\dim(W \cap F) \le k - i - 1 - j \quad \forall F \in \mathcal{F}_j, \ 1 \le j \le k - i - 1$, hence also

$\dim(W \cap (F + \ell)) \le k - i - 1 - j \quad \forall F \in \mathcal{F}_{j+1} \text{ and } 1 \le j < k - i - 1$.

Finally, $W \cap (\mathcal{F}_{k-i} + \ell) = \{0\}$ follows from $W \underset{U}{\pitchfork} (\mathcal{F} \cap U) + \ell$.

Summarizing, we obtain

$$
Y_0^{<W} = \{0 < A < W \mid A \underset{W}{\pitchfork} \mathcal{E}' \text{ and } A \underset{V}{@} \mathcal{F}'_j \text{ for } \dim A = j\} = X_{\mathcal{E}', \mathcal{F}'}(W; V) ,
$$

where $s' := \# \mathcal{F}' \le 2s$ and $\displaystyle\sum_{j=1}^{k-i-1} \binom{k-i-2}{j-1} e'_j + 2^{k-i-1} s' \le \# K$.

4. The poset $Y^{>W}$

We just require $W \in Y$ and $\dim W = k - i < k - 1$ here. Set $\overline{V} := W^\perp/W$, $\overline{U} := U/W$, $\overline{E} := E + W/W$ $(E \in \mathcal{E})$, $\overline{\mathcal{E}} := \{\overline{E} \mid E \in \mathcal{E}\}$, $\overline{F} := (F \cap W^\perp) + W/W$ $(F \in \mathcal{F})$, $\overline{\mathcal{F}}_j := \{\overline{F} \mid F \in \mathcal{F}_{k-i+j}\}$ $(1 \le j \le i - 1)$ and $\overline{\mathcal{F}} := \overline{\mathcal{F}}_1$. Since $\dim(U \cap F) < i$ for all $F \in \mathcal{F}_{k-i+1}$ and $W \underset{U}{\pitchfork} (\mathcal{F} \cap U)$, $W \cap F = 0 = W \cap F^\perp$. Hence $\dim \overline{F} = n - (k - i) =: \overline{n}$ for all $\overline{F} \in \overline{\mathcal{F}}$. Now $\overline{\mathcal{F}} \subseteq \mathcal{U}_{\overline{n}}(\overline{V})$ follows from $W @ \mathcal{F}_{k-i} \supseteq \mathcal{F}_{k-i+1}$. Furthermore, the identity $\overline{U} \cap \overline{F} = (U \cap F) + W/W$ implies $\overline{U} \cap \overline{F} \subseteq \overline{\mathcal{E}} \cup \{0\}$ as well as $\dim(\overline{U} \cap \overline{F}) \leq i - 1 - j$ for all $\overline{F} \in \overline{\mathcal{F}}_j$ $(1 \leq j \leq i - 1)$. Hence $(\overline{V}; \overline{U}; \overline{\mathcal{E}}; \overline{\mathcal{F}}_j, \; 1 \leq j \leq i - 1)$ satisfies all conditions of this lemma (the inequality is easily checked). Observing

$$B \underset{U}{\pitchfork} \mathcal{E} \iff B/W \underset{\overline{U}}{\pitchfork} \overline{\mathcal{E}} \text{ for any } W < B < U \text{ and } B \underset{V}{@} \mathcal{F}_{k-i+j} \iff B/W \underset{\overline{V}}{@} \overline{\mathcal{F}}_j \text{ for}$$

$W < B < U$ and $\dim B = k - i + j$ (cf. Lemma 32 iii)), we thus obtain

$$Y^{>W} \cong \{0 < \overline{B} < \overline{U} \mid \overline{B} \underset{\overline{U}}{\pitchfork} \overline{\mathcal{E}} \text{ and } \overline{B} \underset{\overline{V}}{@} \overline{\mathcal{F}}_j \text{ for } \dim \overline{B} = j\} = X_{\overline{\mathcal{E}};\overline{\mathcal{F}}}(\overline{U}; \overline{V}).$$

5. Two dimension-decreasing filtrations

First we set

$$Z := \{W \in Y \mid \ell \leq W \text{ or } W \underset{U}{\pitchfork} (\mathcal{F} \cap U) + \ell\} \supseteq Y_0 \quad \text{and}$$

$$Y_i := \{W \in Z \mid W \in Y_0 \text{ or } \dim W \geq k - i\}, \; 1 \leq i \leq k - 1 \;.$$

Step 3 shows $Y_0^{<W} = X_{\mathcal{E}';\mathcal{F}'}(W; V)$ for $W \in Y_i \setminus Y_{i-1}$ (and $i < k - 1$).

Now we assume $i > 1$. $Z^{>W}$ contains

$$R := \{W < B < U \mid B \underset{U}{\pitchfork} \mathcal{E} \cup ((\mathcal{F} \cap U) + \ell) \text{ and } B \underset{V}{@} \mathcal{F}_j \text{ for } \dim B = j\}$$

By Step 4, this poset is isomorphic to $X_{\overline{\mathcal{D}};\overline{\mathcal{F}}}(\overline{U}; \overline{V})$, where $\mathcal{D} := \mathcal{E} \cup ((\mathcal{F} \cap U) + \ell)$. Note that $\overline{s} := \#\overline{\mathcal{F}} \leq s$, $\overline{d}_j := \#\overline{\mathcal{D}}_j \leq e_j + s$ and hence

$$\sum_{j=1}^{i-1} \binom{i-2}{j-1} \overline{d}_j + 2^{i-1}\overline{s} \leq \sum_{j=1}^{i-1} \binom{i-2}{j-1} e_j + 2^{i-2}s + 2^{i-1}s \leq \#K$$

In order to see that Flag $Z^{>W}$ is $(i - 2)$-spherical, we consider the filtration defined by

$$R_h := \{W < B < U \mid B \in R \text{ or } (\dim B \geq k - h \text{ and } B \in Z)\} \; (0 \leq h \leq i - 1)$$

Since $\ell \leq W'$ for any $W' \in R_h \setminus R_{h-1}$, the structure of $Z^{>W'} = Y^{>W'}$ is determined by Step 4. Finally, a further application of Step 4 shows that $R^{<W'}$ is isomorphic to $X_{\overline{\mathcal{D}'};\overline{\mathcal{F}}}(\overline{W'}, \overline{V})$ with $\mathcal{D}' := \mathcal{D} \cap W'$.

We complete the filtration of Y by defining

$$Z_i := \{W \in Y \mid W \in Z \text{ or } \dim W \geq k - i\}, \; 0 \leq i \leq k - 1$$

Given $W \in Z_i \setminus Z_{i-1}$, $Z_{i-1}^{>W} = Y^{>W}$ is well known meanwhile.

Finally, $Z_{i-1}^{<W} = Z^{<W} = X_{\mathcal{D} \cap W;\mathcal{F}}(W; V)$, where again $\mathcal{D} := \mathcal{E} \cup ((\mathcal{F} \cap U) + \ell)$. $\quad\square$

Lemma 34: *Let $U, k, \mathcal{E}$ and $e_j (1 \le j \le k-1)$ be given as in Lemma 33. Assume that $\mathcal{F}$ is a subset of $\mathcal{U}_n(V)$ with $s < \infty$ elements satisfying*

(1) $U \cap \mathcal{F} \subseteq \mathcal{E} \cup \{0\}$ *and*

(2) $U \cap F^\perp = 0$ *and* $\dim(U \cap F) \le 1$ *for all* $F \in \mathcal{F}$

Set $Z_{\mathcal{E};\mathcal{F}}(U;V) := \{0 < W < U \mid W \pitchfork_U \mathcal{E} \text{ and } W \,@_V\, \mathcal{F}\}$. *Then* $\text{Flag } Z_{\mathcal{E};\mathcal{F}}(U;V)$ *is $(k-2)$-spherical if* $\#K \ge \sum_{j=1}^{k-1} \binom{k-2}{j-1} e_j + 2s$.

Proof: This proof is easier than that of Lemma 33 since one can use a hyperplane $H < U$ instead of a line ℓ and apply the trivial part of Lemma 32 iii). Before, the following two statements are deduced similar as above:

(1) There exists a line $\ell \in Z_{\mathcal{E};\mathcal{F}}(U;V) =: Y$
   Here $U \cap F^\perp = 0$ and $k \ge 2$ imply $U \not\le M^\perp$ for all $M \in \mathcal{M}(F)$ and $F \in \mathcal{F}$.
   Having established this, statement (1) follows as in the proof of Lemma 33.

(2) $Y^{>W} \cong Z_{\overline{\mathcal{E}};\overline{\mathcal{F}}}(\overline{U};\overline{V})$ for $W \in Y$, $\overline{V} := W^\perp/W$, $\overline{U} := U/W$,
   $\overline{\mathcal{E}} := \{E + W/W \mid E \in \mathcal{E}\}$ and $\overline{\mathcal{F}} := \{(F \cap W^\perp) + W/W \mid F \in \mathcal{F}\}$
   The verification of this statement is analogous to but easier than Step 4 in the proof of Lemma 33.

Combining (1) and (2), one obtains

(3) There exists an $H < U$ such that $\dim H = k - 1$ and $H \in Y$
   Now we define
   $$Y_0 := \{B \in Y \mid B \cap H \in Y\}$$
   $$= \{B \in Y \mid B \le H \text{ or } (\dim B > 1 \text{ and } B \pitchfork_U (\mathcal{E} \cap H))\}$$
   and set $Y_i := \{W \in Y \mid W \in Y_0 \text{ or } \dim B \le i\}$ $(1 \le i \le k - 1)$

It follows for any $W \in Y_i \setminus Y_{i-1}$:

(4) $Y_{i-1}^{<W} = Y^{<W} = \{0 < A < W \mid A \pitchfork_W \mathcal{E} \cap W\}$

Replacing $\mathcal{E}$ by $\mathcal{E}' := \mathcal{E} \cup (\mathcal{E} \cap H)$ in (2), one also obtains

99

(5) $Y_{i-1}^{\geq W} = Y_0^{\geq W} = \{W < B < U \mid B \underset{U}{\pitchfork} \mathcal{E}' \text{ and } B \underset{V}{@} \mathcal{F}\} \cong Z_{\overline{\mathcal{E}'},\overline{\mathcal{F}}}(\overline{U};\overline{V})$

Thus the induction hypothesis can be applied. □

We are turning to subcomplexes of $\Delta = \text{Flag } X$ at last.

Propostion 14: *Let* $\mathcal{E} = \mathcal{E}^{\perp}$ *be a finite set of subspaces of* V ,
$\mathcal{E}_j := \{E \in \mathcal{E} \mid \dim E = j\}$, $e_j := \#\mathcal{E}_j$, $\mathcal{U}_n$ *as in Lemma 32 and*
$\hat{\mathcal{E}}_n := \mathcal{E}_n \cap \mathcal{U}_n$. *Set* $Y_{\mathcal{E}}(V) := \{U \in X \mid U\tilde{\pitchfork}\mathcal{E} \text{ and } (U @ \hat{\mathcal{E}}_n \text{ for } \dim U < n)\}$. *Then*
Flag $Y_{\mathcal{E}}(V)$ *is* $(n-1)$-*spherical if* $\#K \geq 2 \sum\limits_{j=1}^{2n-1} \binom{2n-2}{j-1} e_j$.

Proof: This time we start the induction with $n = 2$ instead of $n = 1$. Nevertheless, we first establish the following, meanwhile familiar statement:

1. There exists a line $\ell \in Y_{\mathcal{E}}(V) =: Y$

Set $\mathcal{D} := (\mathcal{E} \cup \{M^{\perp} \mid M \in \mathcal{M}(E), E \in \hat{\mathcal{E}}_n\}) \setminus (\{E \in \hat{\mathcal{E}}_n \mid \mathcal{M}(E^{\perp}) \neq \emptyset\} \cup \{0, V\})$
Lemma 32 vi) implies

$$\#\mathcal{D} \leq e_1 + \ldots + e_{n-1} + 2e_n + e_{n+1} + \ldots + e_{2n-1} \leq \frac{1}{2}\#K$$

According to the results of §5 (cf. Lemma 25 and Corollary 14), there exists an isotropic line ℓ in V satisfying $\ell \nleq D$ for all $D \in \mathcal{D}$. Then clearly $\ell\tilde{\pitchfork}\mathcal{E}$, and $\ell @ \hat{\mathcal{E}}_n$ follows from Lemma 32 v).

2. The case $n = 2$

Applying 1., we find a line $\ell \in Y$. By Lemma 32 ii), the two maximal totally isotropic subspaces U_1 and U_2 containing ℓ are also elements of Y . It is now sufficient to show that any $U \in Y \setminus \{U_1, U_2\}$ with $\dim U = 2$ can be connected by a path in Flag Y with that U_i satisfying $U \cap U_i = 0$. We may assume $i = 1$. If we can find a line $g < U$ such that $g \in Y$ and $g_1 := g^{\perp} \cap U_1 \in Y$, the path

$$
\begin{array}{ccccc}
\overset{U}{\bullet} & \overset{g}{\circ} & \overset{g \oplus g_1}{\bullet} & \overset{g_1}{\circ} & \overset{U_1}{\bullet}
\end{array}
$$

provides the desired connection. Now $g \in Y$ is equivalent to $g \nleq D$ for all $D \in \mathcal{D}$, where $\mathcal{D}$ is defined as under 1. Note that $U\tilde{\pitchfork}\mathcal{E}$ implies $U \nleq D$ for all $D \in \mathcal{D}$. As for g_1 , we observe that

$$g_1 \in Y \iff g^{\perp} \cap U_1 \nleq D \quad \forall D \in \mathcal{D} \iff D^{\perp} \nleq g + U_1 \quad \forall D \in \mathcal{D}$$

For all $D \in \mathcal{D}$, $U_1 + D^\perp$ is either equal to V (in which case $D^\perp \not\leq g + U_1$ holds trivially) or else 3-dimensional. In the latter case, one obtains

$D^\perp \not\leq g + U_1 \iff U_1 + D^\perp \neq g + U_1 \iff g \not\leq U_1 + D^\perp$. Now choose $0 < g < U$ such that $g \neq U \cap D$ for all $D \in \mathcal{D}$ and $g \neq U \cap (U_1 + D^\perp)$ for all $D \in \mathcal{D}$ satisfying $\dim(U_1 + D^\perp) = 3$. This is possible since

$$\#K \geq 2(e_1 + 2e_2 + e_3) \geq 2\#\mathcal{D} \ .$$

We assume $n \geq 3$ in the following.

3. Description of Y_0

The definition Y_0 here is motivated by that in the proof of Proposition 13. However, the relation @ and some technical pecularities have to be taken into account. Set

$$\mathcal{F} \ := \ \mathcal{E} \cup (\mathcal{E} \cap \ell^\perp) \cup (\mathcal{E} + \ell) \cup ((\mathcal{E} \cap \ell^\perp) + \ell) \quad \text{and}$$

$$\hat{\mathcal{F}}_n \ := \ \mathcal{F}_n \cap \mathcal{U}_n \ . \quad \text{Then}$$

$$\hat{\mathcal{F}}_n \ = \ \hat{\mathcal{E}}_n \cup (\mathcal{E}_{n+1} \cap \ell^\perp) \cup (\mathcal{E}_{n-1} + \ell) \cup ((\hat{\mathcal{E}}_n \cap \ell^\perp) + \ell)$$

since $\ell \pitchfork \mathcal{E}$ and ℓ @ $\hat{\mathcal{E}}_n$. Consider the following conditions

i) $\ell \leq U$, $U \widetilde{\pitchfork} \mathcal{E}$ and U @ $\hat{\mathcal{E}}_n$ for $\dim U < n$

ii) $\ell \not\leq U \leq \ell^\perp$ and $(U \pitchfork \mathcal{E}, \ U$ @ $\hat{\mathcal{E}}_n$ for $\dim U = n - 1$

$\qquad\qquad\qquad\qquad U \pitchfork \mathcal{E} \cup (\mathcal{E} + \ell), \ U$ @ $\hat{\mathcal{F}}_n$ for $\dim U \leq n - 2)$

iii) $U \not\leq \ell^\perp$, $\dim U > 1$, $U \widetilde{\pitchfork} \mathcal{F}$ and U @ $\hat{\mathcal{F}}_n$ for $\dim U < n$

and define $Y_0 := \{U \in X \mid U$ satisfies i), ii) or iii)$\}$.

It is not difficult to verify $U \cap \ell^\perp \in Y_0$ and $(U \cap \ell^\perp) + \ell \in Y_0$ for all $U \in Y_0$ and thus the contractibility of $|\operatorname{Flag} Y_0|$. One just has to use the following observations:

(1) If $\ell \not\leq U$, then $U + \ell \pitchfork \mathcal{E} \iff U \pitchfork \mathcal{E} + \ell$ (cf. §4)

(2) If $U \leq \ell^\perp$ and $\dim U = n - 1$, $U \in Y$ implies $U + \ell \in Y$

$\qquad$ (cf. Lemma 32)

(3) If $\ell \not\leq U \leq \ell^\perp$ and $\dim U \leq n - 2$, then

$\qquad U + \ell$ @ $\hat{\mathcal{E}}_n \iff U$ @ $(\hat{\mathcal{E}}_n \cap \ell^\perp) + \ell$ (cf. the proof of Lemma 33)

(4) If $U \not\leq \ell^\perp$, then $U \cap \ell^\perp \pitchfork \mathcal{E} \iff U \pitchfork \mathcal{E} \cap \ell^\perp \iff U \widetilde{\pitchfork} \mathcal{E} \cap \ell^\perp$

$\qquad$ (cf. the proof of Proposition 13 and note that $\mathcal{E}_{n+1} \cap \ell^\perp \subseteq \mathcal{U}_n$)

(5) If $\ell \not\leq U \not\leq \ell^\perp$, then $(U \cap \ell^\perp) + \ell \, \tilde{\pitchfork} \, \mathcal{E} \iff U \tilde{\pitchfork} (\mathcal{E} \cap \ell^\perp) + \ell$

(This follows from $\dim((U \cap \ell^\perp) + \ell) \circ E = \dim U \circ ((E \cap \ell^\perp) + \ell)$ for $\circ = +$ or $\cap$ and $E \in \mathcal{E}$)

(6) If $U \not\leq \ell^\perp$ and U is maximal totally isotropic, then $U \pitchfork E$ together with $(U \cap \ell^\perp) + \ell \pitchfork E$ implies $U \cap \ell^\perp \, @ \, E$ for any $E \in \hat{\mathcal{E}}_n$

(if the line $(U + \ell) \cap E$ were isotropic, the maximal totally isotropic subspace $(U \cap \ell^\perp) + ((U + \ell) \cap E)$ had to coincide with $(U \cap \ell^\perp) + \ell$ contrary to the assumption $(U \cap \ell^\perp) + \ell \pitchfork E)$

4. A dimension-increasing filtration

We introduce a further condition:

iv) $U \not\leq \ell^\perp$ and $(U \tilde{\pitchfork} \mathcal{F}$ for $\dim U = n$

$\qquad\qquad\qquad U \pitchfork \mathcal{F}$ and $U \, @ \, \hat{\mathcal{F}}_n$ for $\dim U = n - 1$

$\qquad\qquad\qquad U \pitchfork \mathcal{E} \cup (\mathcal{E} + \ell)$ and $U \, @ \, \hat{\mathcal{F}}_n$ for $\dim U \leq n - 2)$

Set $Z \ := \{U \in X \,|\, U$ satisfies i), ii) or iv$)\}$ and

$\qquad Y_i \ := \{U \in Z \,|\, U \in Y_0$ or $\dim U \leq i\}$, $1 \leq i \leq n - 2$.

Let $U \in Y_i \setminus Y_{i-1}$ be given. In particular, $U \not\leq \ell^\perp$ and $\dim U = i \leq n - 2$.

a) Note that $U \, @ \, \hat{\mathcal{F}}_n$ implies $W \, @ \, \hat{\mathcal{F}}_n$ for any $W < U$. Hence we obtain as in the proof of Proposition 13 $Z^{<U} = \{0 < W < U \,|\, W \underset{U}{\pitchfork} \mathcal{E}'\}$, where

$\mathcal{E}' := (\mathcal{E} \cap U) \cup ((\mathcal{E} + \ell) \cap U)$ and $\sum_{j=1}^{i-1} \binom{i-2}{j-1} e'_j \leq \#K$.

b) Set again $\overline{V} := U^\perp / U$, $\overline{n} := n - i$ and

$\overline{\mathcal{E}} := \{\overline{F} := (F \cap U^\perp) + U / U \,|\, F \in \mathcal{F}\}$. Then we obtain for any $W \in X^{>U}$:

(7) $W \underset{V}{\pitchfork} F \iff W/U \underset{\overline{V}}{\pitchfork} \overline{F}$ for arbitrary $F \in \mathcal{F}$

This was already shown in Step 3 b) of the proof of Proposition 13.

(8) Since $U \pitchfork \mathcal{E} \cup (\mathcal{E} + \ell)$, $U \pitchfork \mathcal{F}_n$ as well.

(9) $\dim F = n \iff \dim \overline{F} = \overline{n}$ for any $F \in \mathcal{F}$

"$\Rightarrow$" follows from (8) and $\mathcal{F}_n = \mathcal{F}_n^\perp$. For the reverse direction, we observe that

$1 < \bar{n} < 2\bar{n} - 1$ and the points (2) – (5) in the proof of Proposition 13 imply $U^\perp + F = V$ and $U \cap F = 0$.

Hence $\bar{n} = \dim \overline{F} = \dim F - i$.

(10) $\widetilde{\mathcal{E}}_{\bar{n}} = \{\overline{F} \mid F \in \mathcal{F}_n\}$ and $\widehat{\widetilde{\mathcal{E}}}_{\bar{n}} = \{\overline{F} \mid F \in \widehat{\mathcal{F}}_n\}$

This follows from (9) and $U \text{ @ } \widehat{\mathcal{F}}_n$.

(7) – (10) together with Lemma 32 now imply

(11) $W \stackrel{\sim}{\underset{V}{\pitchfork}} \mathcal{F} \iff W/U \stackrel{\sim}{\underset{V}{\pitchfork}} \widetilde{\mathcal{E}}$

Lemma 32 together with (10) also yields

(12) $W \underset{V}{\text{ @ }} \widehat{\mathcal{F}}_n \iff W/U \underset{V}{\text{ @ }} \widehat{\widetilde{\mathcal{E}}}_{\bar{n}}$ provided that $\dim W < n$

Thus we have proved $Y_0^{>U} \cong Y_{\overline{\mathcal{E}}}(\overline{V})$. Note that by (10) in the proof of Proposition 13 and by Lemma 30, $2 \sum_{j=1}^{2\bar{n}-1} \binom{2\bar{n}-2}{j-1} \bar{e}_j \leq \#K$.

5. Description of $Z^{<U}$ for $U \in Y$, $\ell \not\leq U$ and $\dim U \leq n - 1$

Set $\mathcal{E}' := (\mathcal{E} \cap U) \cup ((\mathcal{E} + \ell) \cap U)$. Let $0 < W < U$ be given and suppose $W \underset{U}{\pitchfork} \mathcal{E}'$.
Then it follows:

(13) $W \underset{V}{\pitchfork} \mathcal{E} \cup (\mathcal{E} + \ell)$ (cf. §4, proof of Proposition 12, Step 3)

(14) $W \text{ @ } \mathcal{E}_{n-1} + \ell$ implies $W \text{ @ } \mathcal{E}_{n+1} \cap \ell^\perp$

This follows from $W \underset{V}{\pitchfork} (\mathcal{E}_{n-1} + \ell) \cup (\mathcal{E}_{n+1} \cap \ell^\perp)$ and Lemma 32 iv).

(15) $U \cap F^\perp = 0$ for $F \in \widehat{\mathcal{E}}_n \cup (\mathcal{E}_{n-1} + \ell)$

(16) $U \cap F^\perp = 0$ or $U \text{ @ } F$ for $F = (E \cap \ell^\perp) + \ell$ and $E \in \widehat{\mathcal{E}}_n$

If $U \cap F^\perp \neq 0$, then $U^\perp + F \neq V$ which implies $U^\perp \cap E \cap \ell^\perp = U^\perp \cap E$ by comparing dimensions and using $U^\perp + E = V$. Since $U \text{ @ } E$, it follows that $U^\perp \cap E \cap \ell^\perp$ is not totally isotropic.

Setting $\mathcal{G} := \widehat{\mathcal{E}}_n \cup (\mathcal{E}_{n-1} + \ell) \cup \{F \in (\widehat{\mathcal{E}}_n \cap \ell^\perp) + \ell \mid U \cap F^\perp = 0\}$, one obtains $Z^{<U} = \{0 < W < U \mid W \underset{U}{\pitchfork} \mathcal{E}'$ and $W \text{ @ } \mathcal{G}\} = Z_{\mathcal{E}';\mathcal{G}}(U; V)$, and the conditions of Lemma 34 are easily checked now.

6. On elements $U \in Y$ with $\dim U = n - 1$

The reasoning of Step 4 breaks down here since (9) is not true any longer. Instead of looking for conditions implying the 0-sphericity of $Y_0^{>U}$, we simply require $Y_0^{>U} \neq \emptyset$ in this case und define

$$Y_{n-1} := Z' := Z \cup \{U \in Y \mid \dim U = n - 1 \text{ and } Y_0^{>U} \neq \emptyset\}\ .$$

Note that Flag Z' is $(n - 1)$-spherical because of Step 4, Step 5 and Lemma 34. Preparing the next filtration, we show that Z' contains "enough" elements of dimension $n - 1$.

(17) Let U_1, U_2 be the two maximal totally isotropic subspaces containing a given $(n - 1)$-dimensional $U \in Y$. Suppose $U_1 \widetilde{\pitchfork} \mathcal{E}$, $U \underset{V}{\pitchfork} (\mathcal{F} \cap U_1)$ and $U @ F$ for all $F \in \hat{\mathcal{F}}_n$ satisfying $U_1 \underset{V}{\pitchfork} F$. Then $U_2 \widetilde{\pitchfork} \mathcal{F}$.

Assume $F \in \mathcal{F} = \mathcal{F}^\perp$. If $\dim F > n$, then $\dim(U_1 \cap F) \geq 1$, implying $U_1 + F = U + F$, and $\dim(U_1 \cap F^\perp) \leq 1$ because of $U_1 \widetilde{\pitchfork} \mathcal{E}$, hence $U \cap F^\perp = 0$. Therefore, $V = U^\perp + F = U_2 + U_1 + F = U_2 + F$. If $\dim F < n$, then $\dim F^\perp > n$ and hence $U_2 + F^\perp = V$, $U_2 \cap F = 0$.

Now we assume $\dim F = n$. If $\dim(U_1 \cap F) = 2$, then $F \in X$, $\dim(U \cap F) = 1$, $U + F = U_1 + F$, $U^\perp + F = U_2 + F$ and $\dim(U^\perp + F) = 2n - 1$, implying $U_2 \widetilde{\pitchfork} F$. If $\dim(U_1 \cap F) = 1$, then $\dim(U_1 \cap F^\perp) \neq 2$, hence $\dim(U_1 \cap F^\perp) \leq 1$ and $U_2 + F = V$ as above. Finally, suppose that $U_1 \cap F = 0$ and $F \notin X$ (otherwise $U_2 \widetilde{\pitchfork} F$ is obvious). Then $U^\perp \cap F \not\leq U_2$ since $U @ F$. Therefore $0 = (U^\perp \cap F) \cap U_2 = U_2 \cap F$.

7. A dimension-decreasing filtration

Set $Y_i := \{U \in Y \mid U \in Z' \text{ or } \dim U \geq 2n - i\}$ for $n \leq i \leq 2n - 1$. For given $U \in Y_i \setminus Y_{i-1}$, we set $k := 2n - i = \dim U$ and distinguish the cases $k = n$ and $k < n$ now.

a) For $\dim U = n$, define $\mathcal{E}' := (\mathcal{E} \cap U) \cup ((\mathcal{E} + \ell) \cap U)$, $\mathcal{F}' := \mathcal{F} \cap U$, $\mathcal{H} := \mathcal{H}_1 := \ldots := \mathcal{H}_{n-2} := \hat{\mathcal{F}}_n$ and $\mathcal{H}_{n-1} := \{F \in \hat{\mathcal{F}}_n \mid U \cap F = 0\}$.

(18) If $F \in \{E, E + \ell\}$, $E \in \mathcal{E}$, $W < U$ and $\dim W \leq n - 2$, then
$$W \underset{V}{\pitchfork} F \iff W \underset{U}{\pitchfork} (F \cap U)\ .$$

104

For $E \notin \tilde{\mathcal{E}}_n := \mathcal{E}_n \setminus \hat{\mathcal{E}}_n$, this follows from $U \pitchfork E$ and for $E \in \tilde{\mathcal{E}}_n$ from $\dim(F \cap U) \leq 2$.

(19) $W < U$, $\dim W = n - 1$, $W \underset{U}{\pitchfork} \mathcal{F}'$ and $W \underset{V}{@} \mathcal{H}_{n-1}$ imply $Y_0^{>W} \neq \emptyset$.

This ist just a reformulation of (17).

(20) If $W < U$, $\dim W \leq n - 2$, $W \underset{U}{\pitchfork} \mathcal{E}'$ and $0 \neq F' \in \mathcal{F}'$, then $F' \not\leq W$.

This is true since there exists an $E' \in \mathcal{E}'$ satisfying $F' \leq E'$ and $\dim E' \leq \dim F' + 1$.

It follows from (18) – (20) that $Y_{n-1}^{<U}$ contains

$$X_{\mathcal{E}';\mathcal{H}}(U; V) \cap \{0 < W < U \mid F' \not\leq W \text{ for all } 0 \neq F' \in \mathcal{F}'\} =: S$$

Using Lemma 33, one deduces (either by considering an appropriate subset $X_{\mathcal{E}'';\mathcal{H}}(U; V)$ of S and then "filling up" or by adding the condition "$F' \not\leq W$ for all $F' \in \mathcal{F}' \setminus \{0\}$" throughout the proof of Lemma 33) that Flag S is $(n - 2)$-spherical. Furthermore, any $W \in Y_{n-1}^{<U} \setminus S$ is of dimension $n - 1$, and Flag $Y_{n-1}^{<W} = $ Flag $S^{<W}$ is $(n - 3)$-spherical by Step 5. Hence Flag $Y_{n-1}^{<U}$ is also $(n - 2)$-spherical.

b) For $\dim U < n$, the poset $Y_{i-1}^{<U} = Z^{<U}$ was studied in Step 5. On the other side, Lemma 32 implies $Y_{i-1}^{>U} = Y^{>U} = Y_{\overline{\mathcal{E}}}(\overline{V})$, where $\overline{V} := U^\perp/U$ and $\overline{\mathcal{E}} := \{\overline{E} := (E \cap U^\perp) + U/U \mid E \in \mathcal{E}\}$ this time. The induction proof is complete now. □

Let $\mathcal{E} = \mathcal{E}(a)$ be given as in Corollary 17. Then $\mathcal{E} \cap \mathcal{U}_n = \emptyset$, $Y_{\mathcal{E}}(V) = X_{\mathcal{E}}(V)$, $e_1, \ldots, e_{n-2}, e_{n+2}, \ldots, e_{2n-1} \leq 1, e_n \leq 2$ and $e_{n-1} = e_{n+1} = 0$. Hence Proposition 14 implies

Corollary 18: *If $\#K \geq 2^{2n-1}$, then $\tilde{\Delta}^0(a)$ is $(n - 1)$-spherical for any simplex a of the thick D_n building $\tilde{\Delta}$.* □

§ 8 Summary

For any spherical building Δ , let us denote by $t(\Delta)$ the **"thickness of Δ"**, i.e.

$$t(\Delta) := \min\{\#\,\mathcal{C}(\mathrm{st}_\Delta(a)) \mid a \in \Delta \text{ is a panel}\}$$

The results stated in §4, Corollary 13, §6, Corollary 16, and in §7, Corollary 18, can be summarized in the following way:

Theorem B: *Let Δ be a classical spherical building in the sense of §3.*

$$Assume\ that\ \ t(\Delta) \geq \begin{cases} 2^{n-1} + 1 & if\ \Delta\ is\ of\ type\ A_n \\ 2^{2n-1} + 1 & if\ \Delta\ is\ of\ type\ C_n \\ 2^{2n-1} + 1 & if\ \Delta\ is\ of\ type\ D_n \end{cases}$$

Then $\Delta^0(a)$ is $(n-1)$-spherical for any $a \in \Delta$. □

In view of Remark 10 and the fact that arbitrary A_2 and C_2 buildings are elementary to deal with (cf.§2), one obtains the following

Corollary 19: *Let Δ be an arbitrary building of type A_n, C_n or D_n but not an exceptional C_3 building. Assume that $t(\Delta)$ satisfies the same condition as in Theorem B. Then $\Delta^0(a)$ is $(n-1)$-spherical for any $a \in \Delta$.* □

III Finiteness properties of classical $\mathbb{F}_q$–groups over $\mathbb{F}_q[t]$

In the last part of this book, we determine the "finiteness length" of the S-arithmetic subgroup $\Gamma = \mathcal{G}(\mathbb{F}_q[t])$ of an almost simple classical isotropic $\mathbb{F}_q$-group $\mathcal{G}$ by combining the main results of Chapter I and Chapter II. Here the term "classical" is used in the sense of [BrT1],§10, i.e. we assume that $\mathcal{G}$ is symplectic, orthogonal, unitary or special linear. It is easy to pass on to groups which are strictly $\mathbb{F}_q$-isogenous to these classical groups afterwards (cf. Corollary 20 below). The exceptional groups are excluded here only because we did not derive a result analogous to Theorem B in these cases.

In order to apply Theorem A, we have to establish the existence of a twin BN-pair in $\mathcal{G}(\mathbb{F}_q[t, t^{-1}])^+$ as claimed in Chapter I,§1, Example 4. If $\mathcal{G}$ splits over $\mathbb{F}_q$, this twin BN-pair — or more precisely: the underlying RGD-system — is described in detail in Example 3. It is also not difficult to introduce an RGD-system in $\mathcal{G}(\mathbb{F}_q[t, t^{-1}])^+$ if $\mathcal{G}$ is a non-split classical $\mathbb{F}_q$-group, i.e. if $\mathcal{G}$ is either non-split orthogonal or unitary. However, this is not described in the literature and will therefore be discussed below, indicating at the same time how to construct an RGD-system in an arbitrary, almost simple isotropic k-group over $k[t, t^{-1}]$.

§ 1 Twin BN-pairs in isometry groups over $k[t, t^{-1}]$

We use again the notations introduced in§5 of Chapter II. For our present purpose, it is sufficient to restrict to the case where K is a field. We denote by k the fixed field of σ . Note that $[K : k] = 2$ and $k = K_{\sigma, \varepsilon} = K^{\sigma, \varepsilon}$ if $\sigma \neq \mathrm{id}_K$. The pseudo-quadratic space V is supposed to be of finite dimension m, of Witt index $n \geq 1$ and to satisfy $V^{\perp} \cap Q^{-1}(0) = 0$. If $\sigma = \mathrm{id}_K$, we only admit the form parameters $\Lambda = 0$ and $\Lambda = K$ corresponding to ordinary orthogonal and to symplectic groups, respectively. Furthermore, we assume $(m, \Lambda) \neq (2n, 0)$ throughout this section because the D_n case has some special features and a twin BN-pair in $SO_{2n}(k[t, t^{-1}])^+$ was already constructed in Example 3.

Fix $2n$ vectors $e_i\, (i \in \{\pm 1, \ldots, \pm n\})$ satisfying $f(e_i, e_j) = \delta_{j, -i}$ for $i > 0$, respectively $f(e_i, e_j) = \varepsilon\, \delta_{j, -i}$ for $i < 0$. Set $V_i := e_i K$ and $V_0 := \bigcap\limits_{i=-n}^{n} e_i^{\perp}$. The isometry group

$$Is(V) := \{ g \in GL(V) \mid f(g(x), g(y)) = f(x, y), \ Q(g(x)) = Q(x) \ \forall x, y \in V \}$$

can be considered as an algebraic k-group (cf. for example [Bo], §23). We define $\mathcal{G} := Is(V) \cap SL(V)$ and

$$\mathcal{T} := \{ g \in \mathcal{G} \mid g(V_i) = V_i \text{ for all } i \in \{0, \pm 1, \dots, \pm n\} \} \ .$$

Note that $\mathcal{T}$ is the centralizer of a maximal k-split torus in $\mathcal{G}$. The system of k-root groups corresponding to that torus is described in detail in [BrT1],§10.1. I will not reproduce all the definitions here but just remind the reader of the following facts:

- Given two indices $i, j \in \{\pm 1, \dots, \pm n\} =: J$ with $i \neq \pm j$, one defines unipotent isometries $u_{ij}(\lambda)$ $(\lambda \in K)$ and monomial matrices $m_{ij}(\lambda)$ $(\lambda \in K^*)$.

- Set $Z := \{ (z, c) \mid z \in V_0, c \in K \text{ and } Q(z) = c + \Lambda \}$. To any $i \in J$ and $(z, c) \in Z$ one associates another unipotent isometry $u_i(z, c)$ and, provided that $(z, c) \neq (0, 0)$, an element $m_i(z, c) \in \mathcal{G}$.

- For fixed (i, j) with $i \neq \pm j$, the elements of the form $u_{ij}(\lambda)$ constitute an abelian root group called $\mathcal{U}_{ij}$ which is k-isomorphic to a $\dim_k K$-dimensional affine space. For $i \in J$, one introduces a metabelian root group $\mathcal{U}_i$ consisting of all elements of the form $u_i(z, c)$. As algebraic group, $\mathcal{G}$ is generated by $\{ \mathcal{U}_i \mid i \in J \} \cup \{ \mathcal{U}_{ij} \mid i, j \in J, \ i \neq \pm j \}$. For any k-algebra A , we denote by $\mathcal{G}(A)^+$ the subgroup of $\mathcal{G}(A)$ generated by all $\mathcal{U}_i(A)$ and all $\mathcal{U}_{ij}(A)$.

The isometries $u_{ij}(\lambda)$, $m_{ij}(\lambda)$, $u_i(z, c)$, $m_i(z, c)$ satisfy various relations, some of them being listed below. Most of these relations appear in [BrT1],§10.1, and all are verified by straightforward calculations. The necessary assumptions (e.g. $i \neq \pm j$ when speaking of $u_{ij}(\lambda)$) are tacitly understood. The sign $\pm$ is often not specified if it is of no importance in the present context.

(1) $u_{ij}(\lambda)^{-1} = u_{ij}(-\lambda)$, $m_{ij}(\lambda)^{-1} = m_{ij}(-\lambda)$

(2) $u_{ji}(\lambda) = u_{ij}(-\varepsilon\lambda^\sigma)$, $m_{ji}(\lambda) = m_{ij}(-\varepsilon\lambda^\sigma)$

(3) $u_i(z, c)^{-1} = u_i(-z, \varepsilon c^\sigma)$, $m_i(z, c)^{-1} = m_i(-z, \varepsilon c^\sigma)$

(4) $m_{ij}(\lambda) = u_{-i,-j}(\pm(\lambda^\sigma)^{-1}) u_{ij}(\lambda) u_{-i,-j}(\pm(\lambda^\sigma)^{-1})$

(5) $m_i(z, c) = u_{-i}(\pm zc^{-1}, \pm(c^\sigma)^{-1})u_i(z, c)u_{-i}(\pm zc^{-1}, \pm(c^\sigma)^{-1})$

(6) $[u_{ij}(\lambda), u_{rs}(\mu)] = 1$ for $\{\pm i, \pm j\} \cap \{\pm r, \pm s\} = \emptyset$

(7) $[u_{ij}(\lambda), u_{js}(\mu)] = 1$ for $s \neq -i$

(8) $[u_{ij}(\lambda), u_{-js}(\mu)] = u_{is}(\pm\lambda\mu)$

(9) $[u_{ij}(\lambda), u_{-ji}(\mu)] = u_i(0, \pm\lambda\mu \pm (\lambda\mu)^\sigma)$

(10) $[u_i(z, c), u_{rs}(\mu)] = 1$ for $i \notin \{-r, -s\}$

(11) $[u_i(z, c), u_{-is}(\mu)] = u_s(\pm z\mu, \pm\mu^\sigma c^\sigma\mu)u_{is}(\pm c\mu)$

(12) $[u_i(z, c), u_j(y, b)] = u_{ij}(f(z, y))$

(13) $[u_i(z, c), u_i(y, b)] = u_i(0, f(z, y) - f(y, z))$

I add some typical conjugation formulae of the form "$mum^{-1} = u'$":

(14) $m_{ij}(\lambda)u_{is}(\mu)m_{ij}(\lambda)^{-1} = u_{-js}(\pm\lambda^{-1}\mu)$

(15) $m_{ij}(\lambda)u_{-i-j}(\mu)m_{ij}(\lambda)^{-1} = u_{ij}(\pm\lambda\mu^\sigma\lambda)$

(16) $m_{ij}(\lambda)u_i(z, c)m_{ij}(\lambda)^{-1} = u_{-j}(\pm z(\lambda^\sigma)^{-1}, \lambda^{-1}c(\lambda^\sigma)^{-1})$

(17) $m_i(z, c)u_{-is}(\mu)m_i(z, c)^{-1} = u_{is}(\pm c\mu)$

(18) $m_i(z, c)u_i(y, b)m_i(z, c)^{-1} = u_{-i}(\pm y'c^{-1}, (c^\sigma)^{-1}bc^{-1})$,
where $y' := m_i(z, c)(y) \in V_0$

By the way, $m_i^0(z, c) := m_i(z, c)|_{V_0}$ is a (quasi-) reflection and in particular

(19) $m_i^0(z\lambda, \lambda^\sigma c\lambda) = m_i^0(z, c)$

For $d \in \mathcal{T}$, we denote by $d_i (i \in J \cup \{0\})$ the ith diagonal entry of d $(d_0 \in Is(V_0))$. We write $d = d_{ij}(\lambda, \mu)$ if $d_i = \lambda$ (and hence $d_{-i} = (\lambda^\sigma)^{-1}$ for $i \neq 0$), $d_j = \mu$ and $d_\ell = 1$ for all $\ell \notin \{\pm i, \pm j\}$. Then it holds:

(20) $d\, u_{ij}(\lambda)d^{-1} = u_{ij}(d_{-i}\lambda d_j^{-1})$

(21) $d\, u_i(z, c)d^{-1} = u_i(d_0(z)d_i^{-1}, d_{-i}cd_i^{-1})$

109

(22) $m_{ij}(\lambda)m_{ij}(\mu) = d_{ij}(-(\mu\lambda^{-1})^\sigma, -\lambda^{-1}\mu)$

(23) $m_i(z,c)m_i(y,b) = d_{0i}(m_i^0(z,c)m_i^0(y,b), \varepsilon(c^\sigma)^{-1}b)$

If we extend the scalars by tensoring with the rational function field $k' := k(t)$, there is an obvious extension (again denoted by σ) of σ to $K' := K(t)$ satisfying $t^\sigma = t$. Then $V' := V \otimes_K K'$ is a (σ, ε)-quadratic K'-space, again of Witt index n , and $\mathcal{G}(k')$ can be identified with the group of isometries of V' with determinant 1. All the relations above remain true if we admit $\lambda, \mu \in K'$ and so on. Analogously, we set $R := k[t, t^{-1}]$, $R' := K[t, t^{-1}]$ and may consider $\mathcal{G}(R)$ as the special isometry group of the (σ, ε)-quadratic R'-module $V \otimes_K R'$.

Now we have to connect group theory with geometry again. Let $\varepsilon_1, \ldots, \varepsilon_n$ be the canonical basis of the Euclidean space $\mathbb{R}^n$. We associate a reduced irreducible root system Ψ and a possibly non-reduced root system $\tilde{\Psi}$ to $\mathcal{G}$ in the following way:

a) If $\mathcal{G}$ is orthogonal, we set (recall that $m \neq 2n$)
$\Psi := \tilde{\Psi} := \{\pm\varepsilon_i \pm \varepsilon_j \mid 1 \leq i \neq j \leq n\} \cup \{\pm\varepsilon_i \mid 1 \leq i \leq n\}$

b) If $\mathcal{G}$ is symplectic or unitary and $m = 2n$, we set
$\Psi := \tilde{\Psi} := \{\pm\varepsilon_i \pm \varepsilon_j \mid 1 \leq i \neq j \leq n\} \cup \{\pm 2\varepsilon_i \mid 1 \leq i \leq n\}$

c) If $\mathcal{G}$ is unitary and $m > 2n$, we define Ψ as under b) and set
$\tilde{\Psi} := \Psi \cup \{\pm\varepsilon_i \mid 1 \leq i \leq n\}$

$\tilde{\Psi}$ is (isomorphic to) the relative root system associated to the reductive k-group $\mathcal{G}$. Setting $\varepsilon_i := -\varepsilon_{-i}$ for $i \in \{-1, \ldots, -n\}$ and defining

$$\mathcal{U}_{\varepsilon_i + \varepsilon_j} \ := \ \mathcal{U}_{ij} \ ,$$
$$\mathcal{U}_{\varepsilon_i} \quad := \ \mathcal{U}_i \quad \text{if} \quad \varepsilon_i \in \tilde{\Psi} \quad \text{and}$$
$$\mathcal{U}_{2\varepsilon_i} \quad := \ \{u \in \mathcal{U}_i \mid u \text{ is of the form } u_i(0, c)\} \text{ if } 2\varepsilon_i \in \tilde{\Psi} \ ,$$

$(\mathcal{T}(k), (\mathcal{U}_a(k))_{a \in \tilde{\Psi}})$ yields a root datum in $\mathcal{G}(k)$ with finite Weyl group
$^vW := W(\tilde{\Psi}) = W(\Psi)$ (cf. [BrT1],§6.1 and§10.1). It is the purpose of the present section to construct an RGD-system with affine Weyl group $W := W_{\text{aff}}(\Psi)$ in $\mathcal{G}(k[t, t^{-1}])^+$. Recall the notions and notations (e.g. $\Pi, \Psi_+, L_{a,\ell}, \mathcal{H}, \alpha_{a,\ell}, \Phi, \Phi_+, S, \Sigma(W, S)$) introduced before the Examples 2 and 3 in Chapter I, 1. We are now prepared for the following

110

Definition 13: *In the context described above, we define the following groups:*

$$G := \mathcal{G}(k[t, t^{-1}])^+ , \quad H := T(k) \cap \mathcal{G}(k)^+$$

and for any $\alpha = \alpha_{a,\ell} \in \Phi$ $(a \in \Psi, \ell \in \mathbb{Z})$

$$U_\alpha := \{u_{ij}(\lambda t^{-\ell}) \,|\, \lambda \in K\} \text{ if } a = \varepsilon_i + \varepsilon_j$$

$$U_\alpha := \{u_i(zt^{-\ell}, ct^{-2\ell}) \,|\, z \in V_0, \ c = Q(z)\} \text{ if } a = \varepsilon_i$$

$$U_\alpha := \{u_i(0, ct^{-\ell}) \,|\, c \in k\} \text{ if } a = 2\varepsilon_i \text{ and } \ell \text{ is odd}$$

$$U_\alpha := \{u_i(zt^{-\frac{\ell}{2}}, ct^{-\ell}) \,|\, (z, c) \in Z\} \text{ if } a = 2\varepsilon_i \text{ and } \ell \text{ is even}$$

I mention in passing that $\mathcal{G}(k) = \mathcal{G}(k)^+$ and $\mathcal{G}(k[t, t^{-1}]) = \mathcal{G}(k[t, t^{-1}])^+$ if $\mathcal{G}$ is symplectic or unitary (cf. for example [HO], Sections 5.3, 6.4 and 9.2 B).

Lemma 35: *The system* $(G, (U_\alpha)_{\alpha \in \Phi}, H)$ *introduced in Definition 13 constitutes an RGD-system with Coxeter group* $W = W_{\mathrm{aff}}(\Psi)$.

Proof: In view of the relations (20) and (21), H normalizes each U_α $(\alpha \in \Phi)$. (RGD0) is clear from the definitions. (RGD1) follows from the commutator formulae (6) – (13) together with

$$[\alpha_{a,\ell}, \alpha_{b,p}] = \{\alpha_{ra+sb, r\ell+sp} \,|\, r, s \geq 0, \ ra + sb \in \Psi \text{ and } r\ell + sp \in \mathbb{Z}\}$$

for $a, b \in \Psi$ with $a \neq -b$ and $\ell, p \in \mathbb{Z}$. In order to establish (RGD2), we first define $m(u_{ij}(\lambda)) := m_{ij}(\lambda)$ and $m(u_i(z, c)) := m_i(z, c)$. Then for any $\alpha = \alpha_{a,\ell} \in \Phi$ and any $u \in U_\alpha \setminus \{1\}$, the relations (4) and (5) show that $m(u) \in U_{-\alpha} u U_{-\alpha}$ (recall that $-\alpha = \alpha_{-a,-\ell}$). Furthermore, the conjugation formulae (14) – (18) (and variants thereof, using for example (2)) together with the identity $s_{a,\ell}(\alpha_{b,p}) = \alpha_{s_a(b), p-\ell(b,a)}$, where $\langle b, a \rangle := \frac{2(b,a)}{(a,a)}$ imply $m(u)U_\beta m(u)^{-1} = U_{s_{a,\ell}(\beta)}$ for any $\beta \in \Phi$. Finally, the relations (1),(3),(22),(23) and (19) show $m(u)^{-1}m(u') \in H$ for any two $u, u' \in U_\alpha \setminus \{1\}$. As in Example 3, we note that $HU_+ \cap U_- \subseteq \mathcal{G}(k[t^{-1}]) \cap \mathcal{G}(k[t]) = \mathcal{G}(k)$. From $T(k)\mathcal{U}_+(k) \cap \mathcal{U}_-(k) = \{1\}$ it now follows that $HU_+ \cap U_- = (HU_+ \cap \mathcal{G}(k)) \cap (U_- \cap \mathcal{G}(k)) = \{1\}$, verifying (RGD3). Axiom (RGD4) is satisfied by the definition of G . $\qquad\square$

Remark 16: According to Proposition 1 and Example 6 in Chapter I, there is firstly a twin BN-pair and secondly a twin building $\Delta = (\Delta_+, \Delta_-)$ associated to the

RGD-system constructed above. Δ_+ and Δ_- are affine buildings with apartments isomorphic to $\Sigma(W, S)$. For $\varepsilon \in \{+, -\}$, Δ_ε can also be described as the Bruhat-Tits building associated to the valuation of the root datum $(\mathcal{U}_a(k(t)))_{a \in \tilde{\Psi}}$ induced by the discrete valuation ω_ε of $k(t)$ which is trivial on k and satisfies $\omega_\varepsilon(t^{-\varepsilon}) = 1$. If $\tilde{\Psi}$ is of type B_n, C_n, BC_n, respectively, then the "échelonnage" of this Bruhat-Tits building is of type $B_n, C_n, C - BC_n^{IV}$, respectively (cf. [BrT1], 10.1.26).

§ 2   The finiteness length of $\mathcal{G}(\mathbb{F}_q[t])$

In this last section, we assume $k = \mathbb{F}_q$. If $\mathcal{G}$ is as described in §1, then either $K = k = \mathbb{F}_q$ or else $\mathcal{G}$ is unitary and $K = \mathbb{F}_{q^2}$. Applying Theorem A and Theorem B, it is now easy to deduce the following

Theorem C: *Let $\mathcal{G}$ be an absolutely almost simple classical $\mathbb{F}_q$-group of $\mathbb{F}_q$-rank $n \geq 1$. Suppose that q satisfies*

$$q \geq 2^{n-1} \quad \text{if } \mathcal{G} = SL_{n+1} \ ,$$
$$q \geq 2^{2n-2} \quad \text{if } \mathcal{G} = Sp_{2n} \quad \text{and}$$
$$q \geq 2^{2n-1} \quad \text{if } \mathcal{G} = SO_{2n}, \ SO_{2n+1}, \ {}^2SO_{2n+2}, \ SU_{2n} \text{ or } SU_{2n+1}.$$

Then $\mathcal{G}(\mathbb{F}_q[t])$ is of type F_{n-1} but not of type FP_n. Under the same conditions, $\mathcal{G}(\mathbb{F}_q[t, t^{-1}])$ is also of type F_{n-1}.

Proof: We set again $G := \mathcal{G}(\mathbb{F}_q[t, t^{-1}])^+$ and denote by $(G, (U_\alpha)_{\alpha \in \Phi}, H)$ the RGD-system established in Example 3 of Chapter I, respectively in Lemma 35.

Let $(G, B_+ = HU_+, \ B_- = HU_-, \ N, \ S)$ be the twin BN-pair corresponding to this RGD-system (cf. Proposition 1) and $\Delta = (\Delta_+, \Delta_-)$ the twin building associated to it (cf. Example 6). The maximal parabolic subgroup $P_- := B_- W(\Psi) B_-$ corresponds to a vertex of type 0 in Δ_- and is at the same time the stabilizer in G of this vertex. Obviously $P_- \subseteq \mathcal{G}(\mathbb{F}_q[t])^+ \subset G$, implying $P_- = \mathcal{G}(\mathbb{F}_q[t])^+$. In order to apply Theorem A to P_-, one has to verify the conditions (LF), (F) and (S). (LF) is clear since the buildings Δ_+, Δ_- are affine and of irreducible type. (F) follows from the finiteness of $\mathbb{F}_q$ (cf. Lemma 15 and Corollary 7 in Chapter I, §6).

Now let Θ be a proper link in Δ_+. Then Θ is a finite building of rank $\leq n$, and the thickness of Θ in the sense of Chapter II, §8, is $\geq q + 1$ because each of the root groups U_α contains at least q elements. Considering the Coxeter diagrams of affine Weyl groups of type $\tilde{A}_n, \tilde{B}_n, \tilde{C}_n$ and $\widetilde{D}_n$, one sees that Θ is a join of buildings of type A_i, C_j, D_ℓ with $i, j, \ell \leq n$. Now Theorem B implies that $\Theta^0(a)$ is $(\dim \Theta)$-spherical for any $a \in \Theta$ (in case $\mathcal{G} = Sp_{2n}$, one has to apply Corollary 16 of Chapter II, §6, directly). This establishes condition (S).

So we have shown that P_- is of type F_{n-1} but not of type FP_n and also that G is of type F_{n-1} (cf. Corollary 11 in Chapter I, §6, for the last statement). It follows from the K-theory of classical groups (cf. again [HO], Sections 5.3, 6.4, 7.2 and 9.2 B) or from the theory of S-arithmetic groups that $\mathcal{G}(R)^+$ is of finite index in $\mathcal{G}(R)$ if $R = \mathbb{F}_q[t]$ or $\mathbb{F}_q[t, t^{-1}]$. Therefore, $\mathcal{G}(R)$ and $\mathcal{G}(R)^+$ possess the same finiteness properties. $\qquad\qquad\square$

If $\mathcal{G}$ is strictly $\mathbb{F}_q$-isogenous to one of the groups listed in Theorem C, the implications stated there remain valid. This follows from the fact that S-arithmetic groups are mapped onto S-arithmetic groups by strict isogenies (cf. [M], Ch. I, Corollary 3.2.9). Hence we obtain the following

Corollary 20: *Let $\mathcal{G}$ be an arbitrary absolutely almost simple $\mathbb{F}_q$-group which is not of exceptional type. Denote by n the $\mathbb{F}_q$-rank of $\mathcal{G}$. Suppose $n \geq 1$ and $q \geq 2^{2n-1}$.*

Then $\mathcal{G}(\mathbb{F}_q[t])$ and $\mathcal{G}(\mathbb{F}_q[t, t^{-1}])$ are of type F_{n-1} , and $\mathcal{G}(\mathbb{F}_q[t])$ is not of type FP_n. $\qquad\qquad\square$

I will finish by commenting on some aspects of Theorem C.

Remark 17:

i) As far as the properties F_1 and F_2 are concerned, the assumptions regarding q are superfluous. In fact, the finitely generated and finitely presented S-arithmetic subgroups of reductive groups defined over a global function field are completely characterized meanwhile by the work of Behr (cf. [Be1] and [Be2]).

ii) The finiteness length of $SL_{n+1}(\mathbb{F}_q[t])$ for $q \geq 2^{n-1}$ was determined independently by Abels and the author some years ago (cf. [A2] and [Ab1]). In [Ab1], a different filtration adapted to the specific properites of $\tilde{A}_n$ buildings was used, admitting a proof under the less restrictive condition $q \geq \binom{n-1}{[\frac{n-1}{2}]}$.

iii) Presupposing Theorem B, that part of Theorem C dealing with classical Chevalley groups was derived in [Ab3] by using the action of $\mathcal{G}(\mathbb{F}_q[t])$ on the corresponding Bruhat-Tits building but not yet the action of $\mathcal{G}(\mathbb{F}_q[t,t^{-1}])$ on the associated twin building.

iv) If $\mathcal{G}$ is a semisimple $\mathbb{F}_q$-group such that all its almost simple factors are classical, then Theorem C implies analogous results concerning $\mathcal{G}(\mathbb{F}_q[t])$ and $\mathcal{G}(\mathbb{F}_q[t,t^{-1}])$. This is due to the fact that a direct product $\Gamma = \Gamma_1 \times \ldots \times \Gamma_r$ is of type FP_m if and only if all factors Γ_i are.

v) Finally, I want to remind the reader of Remark 7 in Chapter I, §6. The book just completed naturally represents an invitation to determine the exact finiteness length of $\mathcal{G}(\mathbb{F}_q[t,t^{-1}])$ next.

Bibliography

[A1] H. Abels: *Finite Presentability of S-Arithmetic Groups — Compact Presentability of Solvable Groups.* Lecture Notes in Mathematics 1261, Springer (1987).

[A2] H. Abels: *Finiteness properties of certain arithmetic groups in the function field case.* Israel J. Math. **76** (1991), 113 – 128.

[AA] H. Abels and P. Abramenko: *On the homotopy type of subcomplexes of Tits buildings.* Advances in Math. **101** (1993), 78 – 86.

[AB] H. Abels and K.S. Brown: *Finiteness properties of solvable S-arithmetic groups: An example.* J. Pure Appl. Algebra **44** (1987), 77 – 83.

[AH] H. Abels and S. Holz: *Higher generation by subgroups.* J. Algebra **160** (1993), 310 – 341.

[Ab1] P. Abramenko: *Endlichkeitseigenschaften der Gruppen $SL_n(\mathbb{F}_q[t])$.* Dissertation, Frankfurt (1987).

[Ab2] P. Abramenko: *Walls in Coxeter complexes.* Geometriae Dedicata **49** (1994), 71 – 84.

[Ab3] P. Abramenko: *Finiteness properties of Chevalley groups over $\mathbb{F}_q[t]$.* Israel J. Math. **87** (1994), 203 – 223.

[Ab4] P. Abramenko: *Some spherical subcomplexes of spherical buildings I.* Preprint 92–058, SFB Bielefeld (1992).

[Ab5] P. Abramenko: *Group actions on twin buildings.* Bull. Belg. Math. Soc. **3** (1996), 391 – 406 (in press).

[Ab6] P. Abramenko: *A counter-example: Twin buildings of type $(4,4,4)$ over $\mathbb{F}_2$.* Manuscript, Frankfurt (1996).

[AR] P. Abramenko and M. Ronan: *A characterization of twin buildings by twin apartments.* Preprint 93-055, SFB Bielefeld (1993). A modified version is submitted to Geometriae Dedicata.

[Ba] U. Baumgartner: *Eine Verallgemeinerung des Struktursatzes für Gruppenoperationen auf Bäumen mit Fundamentalbereich auf einfach zusammenhängende geordnete Mengen und Anwendungen.* Diplomarbeit, Frankfurt (1992).

[BB] M. Bestvina and N. Brady: *Morse theory and finiteness properties of groups.* Preprint, Salt Lake City (1995).

[Be1] H. Behr: *Finite presentability of arithmetic groups over global function fields.* Proc. Edinburgh Math. Soc. **30** (1987), 23 – 39.

[Be2] H. Behr: *Arithmetic groups over function fields.* Preprint 92–033, SFB Bielefeld (1992).

[Bi] R. Bieri: *Homological dimension of discrete groups.* Queen Mary College Mathematics Notes, London (1976).

[Bo] A. Borel: *Linear Algebraic Groups.* 2nd edition, Springer GTM 126 (1991).

[BoS1] A. Borel and J.-P. Serre: *Corners and arithmetic groups.* Comment. Math. Helv. **48** (1973), 436 – 491.

[BoS2] A. Borel and J.-P. Serre: *Cohomologie d'immeubles et de groupes S-arithmétiques.* Topology **15** (1976), 211 – 232.

[BoT] A. Borel and J. Tits: *Groupes réductifs.* Publ. Math. I.H.E.S. **27** (1965), 55 – 151.

[Bou1] N. Bourbaki: *Algèbre, Chap. IX.* Hermann, Paris (1959).

[Bou2] N. Bourbaki: *Groupes et Algèbres de Lie, Chap. IV – VI.* Hermann, Paris (1968).

[Bous] F. Bourseau: *Transversale Unterkomplexe in Gebäuden.* Dissertation, Bielefeld (1994).

[Brou] A.E. Brouwer: *The complement of a geometric hyperplane in a generalized polygon is usually connected.* In: Finite Geometry and Combinatorics. L.M.S. Lecture Notes 191, Cambridge University Press (1993), 53 – 57.

[Br1] K.S. Brown: *Cohomology of Groups.* Springer GTM 87 (1982).

[Br2] K.S. Brown: *Finiteness properties of groups.* J. Pure Appl. Algebra **44** (1987), 45 – 75.

[Br3] K.S. Brown: *Buildings.* Springer (1989).

[BrT1] F. Bruhat and J. Tits: *Groupes réductifs sur un corps local, I.* Publ. Math. I.H.E.S. **41** (1972), 5 – 251.

[BrT2] F. Bruhat and J. Tits: *Groupes réductifs sur un corps local, II.* Publ. Math. I.H.E.S. **60** (1984), 5 – 184.

[Bu] K.-U. Bux: *Eine Serie metabelscher S-arithmetischer Gruppen.* Diplomarbeit, Frankfurt (1993).

[C1] C. Chevalley: *Sur certaines groupes simples.* Tôhoku Math. J. **7** (1955), 14 – 66.

[C2] C. Chevalley: *Séminaire sur la classification des groupes de Lie algébriques.* Mimeographed notes, Paris (1956 – 58).

[DS] A. Dress and R. Scharlau: *Gated sets in metric spaces.* Aequationes Math. **34** (1987), 112 – 120.

[F] J.R. Faulkner: *Groups with Steinberg relations and coordinatization of polygonal geometries.* Memoirs Am. Math. Soc. vol. 10, no. 185 (1977).

[FS] P. Fong and G.M. Seitz: *Groups with a (B, N)-pair of rank 2, I and II.* Inventiones math. **21** (1973), 1 – 57, and **24** (1974), 191–239.

[HO] A.J. Hahn and O.T. O'Meara: *The Classical Groups and K–Theory.* Springer (1989).

[M] G.A. Margulis: *Discrete Subgroups of Semisimple Lie Groups.* Springer (1991).

[MR] B. Mühlherr and M. Ronan: *Local to global structure in twin buildings.* Inventiones math. **122** (1995), 71 – 81.

[N] H. Nagao: *On $GL(2, K[x])$.* J. Poly. Osaka Univ. **10** (1959), 117 – 121.

[Q] D. Quillen: *Finite generation of the groups K_i of rings of algebraic integers.* In: Algebraic K-Theory I. Springer, Lecture Notes in Mathematics 341 (1973), 179 – 198.

[Ra] M.S. Raghunathan: *A note on quotients of real algebraic groups by arithmetic subgroups.* Inventiones math. **4** (1968), 318 – 335.

[Ro] M. Ronan: *Lectures on Buildings.* Academic Press (1989).

[RT] M. Ronan and J. Tits: *Twin trees I.* Inventiones math. **116** (1994), 463 – 479.

[Sch] R. Scharlau: *Buildings.* In: Handbook of Incidence Geometry. Edited by F. Buekenhout, North Holland, Amsterdam (1995).

[Se1] J.-P. Serre: *Cohomologie des groupes discrets.* Annals of Math. Studies 70, Princeton University Press (1971), 77 – 169.

[Se2] J.-P. Serre: *Trees.* Springer (1980).

[So] C. Soulé: *Chevalley groups over polynomial rings.* In: Homological Group Theory, Proc. Symp. Durham 1977, L.M.S. Lecture Notes 36 (1979), 359 – 367.

[Sp] E.H. Spanier: *Algebraic Topology.* McGraw–Hill (1966).

[St] R. Steinberg: *Lectures on Chevalley Groups.* Mimeographed notes, Yale University (1967).

[Stu] U. Stuhler: *Homological properties of certain arithmetic groups in the function field case.* Inventiones math. **57** (1980), 263 – 281.

[Sw] R.G. Swan: *Generators and relations for certain special linear groups.* Advances in Math. **6** (1971), 1 – 77.

[T1] J. Tits: *Buildings of Spherical Type and Finite BN-Pairs.* Lecture Notes in Mathematics 386, Springer (1974).

[T2] J. Tits: *Classification of buildings of spherical type and Moufang polygons: a survey.* In: Teorie Combinatorie, Proc. Intern. Colloq. Rome 1973, vol. I, Accad. Naz. Lincei (1976), 229 – 246.

[T3] J. Tits: *Endliche Spiegelungsgruppen, die als Weylgruppen auftreten.* Inventiones math. **45** (1977), 283 – 295.

[T4] J. Tits: *A local approach to buildings.* In: The Geometric Vein (Coxeter Festschrift), Springer (1981), 519 – 559.

[T5] J. Tits: *Moufang octagons and the Ree groups of type* $^2F_4$. Amer. J. Math. **105** (1983), 539 – 594.

[T6] J. Tits: *Buildings and group amalgamations.* L.M.S. Lecture Notes 121 (Proceedings of Groups – St. Andrews 1985), Cambridge University Press (1986), 110 – 127.

[T7] J. Tits: *Ensembles ordonnés, immeubles et sommes amalgamées.* Bull. Soc. Math. Belgique **38** (1986), 367 – 387.

[T8] J. Tits: *Uniqueness and presentation of Kac–Moody groups over fields.* J. Algebra **105** (1987), 542 – 573.

[T9] J. Tits: *Résumé de cours.* Annuaire du Collège de France, 89e anné (1988 – 89), 81 – 95.

[T10] J. Tits: *Résumé de cours.* Annuaire du Collège de France, 90e anné (1989 – 90), 87 – 103.

[T11] J. Tits: *Twin buildings and groups of Kac–Moody type.* L.M.S. Lecture Notes 165 (Proceedings of a conference on groups, combinatorics and geometry, Durham 1990), Cambridge University Press (1992), 249 – 286.

[T12] J. Tits: *Moufang Polygons, I. Root data.* Bull. Belg. Math. Soc. **3** (1994), 155 – 168.

[V] K. Vogtmann: *Spherical posets and homology stability for* $O_{n,n}$. Topology **20** (1981), 119 – 132.

Index of notations

Subject Index